Extreme Birds

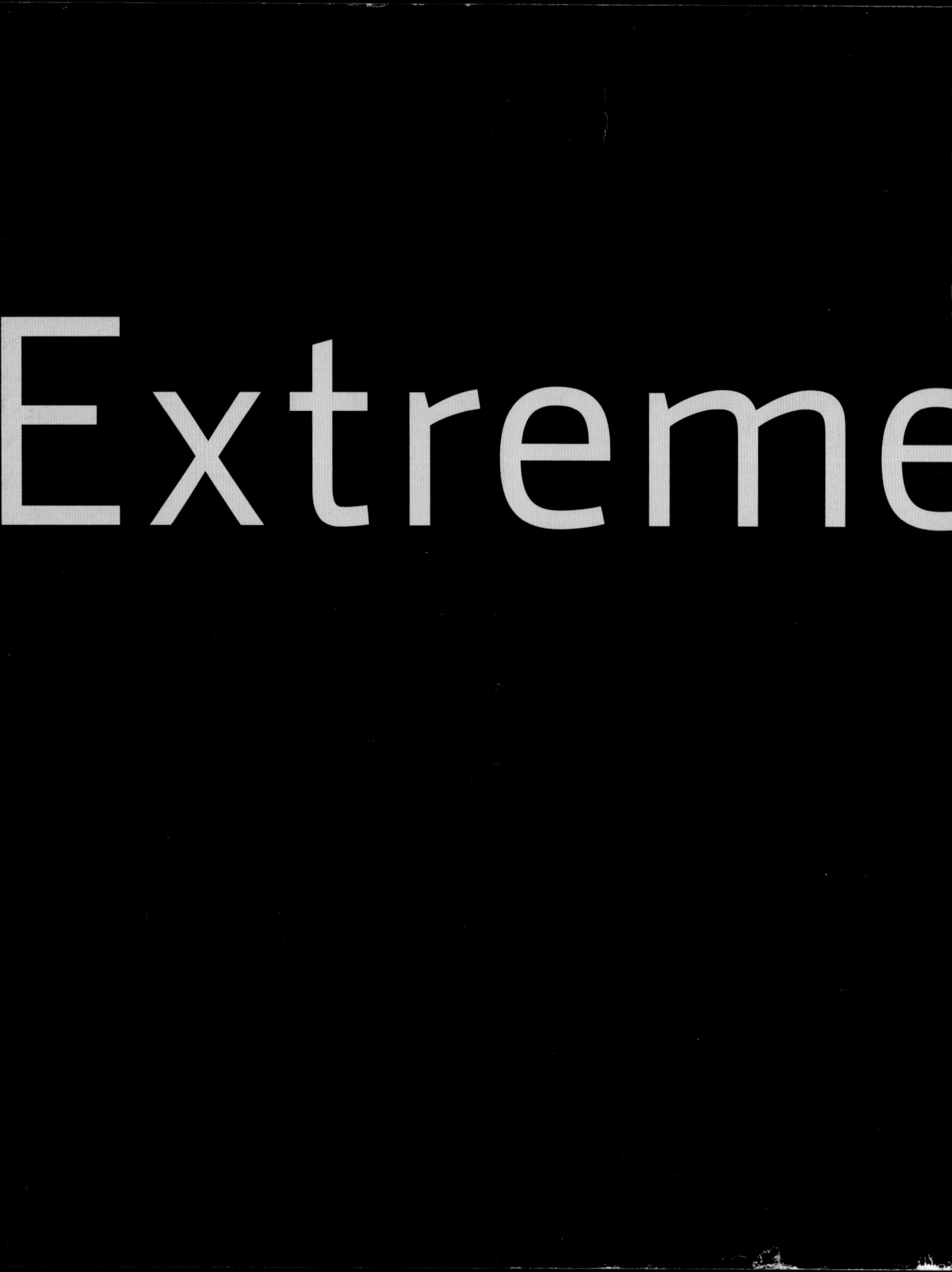
Extreme

Birds

The World's Most Extraordinary and Bizarre Birds

Dominic Couzens

Firefly Books

Published by Firefly Books Ltd. 2011

First printing

Publisher Cataloging-in-Publication Data (U.S.)

Couzens, Dominic.
Extreme birds : the world's most extraordinary and bizarre birds / Dominic Couzens.
[288] p. : col. photos. ; cm.
Includes index.
Summary: Examples of the extremes of the bird world, from biggest and fastest to the smelliest and smartest.
ISBN-13: 978-1-55407-952-0 (pbk.)
1. Birds -- Pictorial works. 2. Birds -- Miscellanea. I. Title.
598 dc22 QL673.C682 2011

Library and Archives Canada Cataloguing in Publication

Couzens, Dominic
Extreme birds : the world's most extraordinary and bizarre birds / Dominic Couzens.
Includes index.
ISBN 978-1-55407-952-0
1. Birds. 2. Birds--Pictorial works. I. Title.
QL673.C69 2011 598 C2011-900961-7

Published in the United States by
Firefly Books (U.S.) Inc.
P.O. Box 1338, Ellicott Station
Buffalo, New York 14205

Published in Canada by
Firefly Books Ltd.
66 Leek Crescent
Richmond Hill, Ontario L4B 1H1

Developed by HarperCollins Publishers Ltd.
77-85 Fulham Palace Road
London W6 8JB

Printed in China

To Matthew and Daniel

Contents

Introduction

Few groups of animals are as visible or abundant as birds. It is difficult to go outside without seeing at least a few, even if they are only pigeons or sparrows. Because of this, birds are perhaps the best known and appreciated of all wildlife.

Many avian record holders are both famous and obvious. The largest mammal might be hidden away under the sea, and the largest snake might keep to itself, but you cannot miss an ostrich wandering over the African savanna. It can't miss you, either, because its eyes are five times as big as yours, the largest of any vertebrate. Meanwhile, the bird with the largest wingspan, the wandering albatross, is a household name and a symbol of the destruction we have wrought in the oceans, while the feats of the fast-flying peregrine falcon and the deep-diving emperor penguin are well known to many enthusiasts.

However, if you delve deeper into the world of birds, you will find some astonishing feats that are not so well known. There is a bird, for example, that can sleep for 100 days, and another that can fly without stopping for a minimum of four years. There are those that can hear in three dimensions and spot a rodent from a mile away, and others that can fly to heights of 30,000 feet (9,000 m) without any side effects. Still others use their mouths as thermometers. Despite the fact that birds are much less physically and physiologically variable than many other animals—as a result of the constraints placed upon them by the need to fly—they still manage to take their body plan and abilities to every possible extreme.

Once we begin to look into the behavior of birds we find even more surprises. Here the "extremes" might not be so obvious, but the ways in which birds live and solve their problems are no less remarkable. Take the Arctic owls that cache the bodies of rodents for later consumption and then defrost them by sitting on them as they would eggs, or the small African parrot that carries its nest material in the feathers of its back, leaving its wings free for flight. The behaviors of birds—those small nuances of difference that have arisen to give a competitive advantage—are a mine of intrigue and astonishment.

Nowhere is this more obvious than during reproduction. Birds, it seems, will go to any lengths to get their genes passed on, be it by rape, deception or parasitism—and sometimes by all three within the same species. Yet while the young may be cosseted inside nests with exceptional insulating properties, they may also be summarily abandoned when conditions for breeding go awry. Young birds don't just sit there and accept what comes to them either; some deliberately kill their siblings, and others may fire lazy parents. Acts of desperate survival are everywhere.

No book of this kind is possible without its source material, so I must thank a group of people whose labors are sometimes overlooked—the researchers. It is they who put in the hard hours of effort and inquiry that may ultimately translate into a single sentence on the page of a book. There is no record without someone to measure it, and no discovery of previously unknown behavior without someone in the field to look and wonder. This book is really a tribute to the researchers' efforts.

It has become fashionable in recent times for every wildlife book to make a plea about conservation, but in this case repetition is apposite, and I make no apology for raking over the obvious once again. A number of the species in this book, including the wandering albatross, the Andean condor, the hooded grebe and the aquatic warbler, have low populations and are close to extinction. They are not included here for their rarity, of course, but for some extreme of lifestyle or behavior. The very fact, however, that every bird book records characteristics that are in danger of being lost is merely symptomatic of how carelessly we have treated our world. I hope that, in its small way, *Extreme Birds* will help fight against this trend and spread a little delight in the feathered creatures with which we share this planet.

Dominic Couzens
Dorset, England, April 2008

Extreme

Form

Widest wingspan · Longest legs · Biggest eyes · Smallest bird · Deadliest enemy · Most effective camouflage · Longest toes · Strongest claws · Biggest mouth · Smelliest bird · Oddest bill · Longest bill · Best energy saver · Most poisonous bird · Snuggest underwear · Ugliest looks · Lightest bones · Heaviest flier · Worst flier · Best submarine · Most flexible mouth · Oddest skin · Longest tongue · Most variable feathers · Heaviest testes · Longest penis · Sexiest tail · Biggest belly · Classiest colors · Whitest bird · Most variation in a species · Best grooming aids · Most useful esophagus

Widest wingspan

NAME **wandering albatross** *Diomedea exulans*
LOCATION southern oceans
ATTRIBUTE longest wings of any bird

A magnificent wandering albatross rides the stormy winds around the island group of South Georgia in the south Atlantic Ocean. With its long, straight wings it can cope with all but the most violent gales, translating the raw energy of tempestuous winds into masterful, effortless flight.

Albatross, of which there are about 20 species, are among the most exciting birds in the world to see, and the wandering albatross is the biggest of all of them. It holds the record for the longest wingspan of any bird, reliably measured at 12 feet (3.6 m; the norm being 8 feet–11½ feet/ 2.5–3.5 m), although informed speculation puts some individuals in excess of 13 feet (4 m).

The wandering albatross looks so powerful that you may be surprised to learn that if it has to flap its wings more than a few times in succession it quickly becomes exhausted. It is a big, heavy bird, and the wings are too narrow to carry its weight in still conditions (technically, it is said to have a high wing loading). However, if a wind of over 11 mph (18 km/h) is blowing, the sheer length of the wings comes into play in positive fashion, generating lift from wing tip to wing tip and almost negligible drag. Hence most albatross are found in the southern oceans, where winds tear around the earth unhindered.

The method used by albatross to travel vast distances with a small expenditure of energy is known as dynamic soaring. The birds glide along the trough of a wave and then suddenly turn into the wind, which scoops them up to a height of 33–50 feet (10–15 m). They then drift down toward the next trough and repeat the process for as long as they need.

A lesser flamingo takes off while the rest of the flock remains stiltlike in the water. Looking at this image, it should be of little surprise to find out that flamingos have the longest legs of any bird relative to its size. A flamingo stands up to 5 feet (1.5 m) tall, while its legs can be up to 27 inches (68 cm) long, accounting for almost half the height of the bird.

Why, though, should flamingos have such long legs? It certainly isn't to help them run, as it is for the ostrich. Being water birds with partly webbed feet, they would soon trip over if they tried to run. The advantage of their long legs is that they help them to wade in deeper water than any competitors. Because flamingos obtain all their food in the same way, by filtering water of microorganisms, they need to be able to work in water of almost any depth. Indeed, occasionally they also swim.

One of the reasons that the legs look so long is that they are absolutely bare, right up to the belly. This is an adaptation to the flamingos' habitat. Flamingos tend to feed in extremely alkaline or saline water, where very few other birds could exist. Some also live near hot springs, where they can wade in water of temperatures up to 154°F (68°C). Such conditions would cause unacceptable wear and tear on the feathers.

Longest legs

NAME **lesser flamingo** *Phoenicopterus minor*
LOCATION Africa
ATTRIBUTE proportionally the longest legs of any bird

Biggest eyes

NAME **ostrich** *Struthio camelus*
LOCATION sub-Saharan Africa
ATTRIBUTE largest eyes of any bird

When ostriches look at you it's hard to avoid their gaze. That's probably because of their rather sinister, long, snakelike neck and serious, unflinching expression. Then again, it could be because they have the largest eyes of any bird in the world.

In fact, this is not the only impressive statistic about an ostrich's eyes. At 2 inches (5 cm) in diameter, measured front to back from the center of the cornea to the retina, they are five times bigger than the human eye and beat that of any land animal; only the mightiest squids in the sea have larger eyes. Furthermore, in a neat comparison of largest to littlest, the eye of an ostrich is about as big as the smallest hummingbird! Altogether it is a remarkable organ.

The ostrich needs large eyes for its terrestrial lifestyle, sharing the savanna as it does with an alarming army of fearsome predators. The bird is famed for its running prowess, reaching speeds of 45 mph (70 km/h) when pressed, which is fast enough to escape most predators. But it has to see them first. Standing up to 9 feet (2.75 m) tall, it enjoys an excellent view over the grassland and bush. Meanwhile, the high number of photoreceptor cells in its eye, combined with the sheer size of the image from the lens, means that the ostrich can see in phenomenal detail. Indeed, its eye is at the size limit of usefulness—any larger and diffraction effects would begin to distort the image.

Smallest bird

NAME	**bee hummingbird**
	Mellisuga helenae
LOCATION	Cuba
ATTRIBUTE	smallest of all birds

The 300 or so species of hummingbird are famous for their diminutive size, so it is not surprising that one of them holds the record for being the smallest bird in the world. That accolade is generally given to the male bee hummingbird, a species from Cuba that is a mere 2¼ inches (5.7 cm) long, but there are plenty of birds that are not far behind. Some of the woodstars (*Chaetocercus* spp.) are only 2½–2¾ inches (6–7 cm) long, and the reddish hermit (*Phaethornis ruber*) rivals the bee hummingbird in lightness, tipping the scales at less than 1/14 ounce (2 g).

For now, however, the bee hummer is officially the smallest, until someone discovers a new hummingbird that has been hitherto overlooked as a flying insect. It could happen—bee hummers really do look like large flying insects, and some species of hummingbird, including the bumblebee hummingbird (*Atthis heloisa*), imitate the flight style of bees in order to avoid eviction from blooms by more aggressive, territorial hummingbirds.

Being the smallest bird in the world also affords the bee hummingbird other records. It is presumed also to have the smallest nest of any bird, measuring only ¾ inch (2 cm) in both diameter and depth. Within this nest it lays what are probably the smallest eggs, measuring just ½ inch (12.5 mm) long by ⅓ inch (8.5 mm) wide and less than half the weight of a paper clip. Indeed, it would take 3,000 bee hummingbird eggs to equal the weight of the world's largest living bird's egg, that of the ostrich.

Deadliest enemy

NAME **southern cassowary**
Casuarius casuarius
LOCATION rain forests of New Guinea and northeast Australia
ATTRIBUTE stabbing and disemboweling people

In April 1926, near the town of Mossman in northeast Queensland, Australia, a group of boys went out hunting cassowaries for fun. As they chased a bird through the rain forest, one of them, a 16-year-old name Phillip McClean, tripped over a branch and fell to the ground. In that brief moment the pursuer became the pursued. The cassowary turned on the boy and slashed him with its central claw, slicing open his jugular vein. The unfortunate McClean thus became the first, and so far only, authenticated fatality from a wild cassowary attack in Australia. He is also one of the very few people ever to have been directly killed by any bird, anywhere in the world.

Cassowaries, of which there are three known species, are large, flightless rain-forest dwellers, the third largest of all living birds after the ostrich and the emu. As tall as an average person, they weigh up to 130 pounds (60 kg), can run at 30 mph (50 km/h) and, according to some reports, can leap 5 feet (1.5 m) into the air. Normally peaceful birds, they rarely attack people except under extreme provocation.

They are well equipped to defend themselves. On top of their head is a thick casque that they use to butt into their enemies, which proves effective against most assailants. However, it's the claws that are truly deadly. A cassowary has three toes per foot, each one bearing a claw. The central claw grows up to 5 inches (12 cm) long and is as sharp as a dagger. Lashing out with it, a bird can easily slice through flesh and disembowel someone. Since 1990 alone, there have been six serious attacks on humans recorded, several of them upon zookeepers, one of which was fatal.

An adult willow ptarmigan takes refuge in a snow burrow, confident that its white plumage will protect it. It is a perfect example of camouflage in a species that, being plump and tasty, is sought out with some vigor by a range of tundra carnivores.

A high proportion of the world's bird species exhibit camouflage of one type or another, but the ptarmigans are exceptional in that their plumage changes up to three times a year to match the seasonal habitat. A white winter coat is suitable until the snow melts, but then it would become a liability, making the birds *more* obvious rather than less. Thus in the spring, prior to breeding, ptarmigans switch plumage. They begin to acquire some reddish brown feathers to match the appearance of the vegetation as it emerges from the snow. At first they are blotchy, as if the white was melting from their feathers, but by midsummer the transformation is complete, and they have become rich brown almost all over. Their cloaking is different, but it is as perfect for the time of year as their winter coat.

After breeding, ptarmigans then have two molts in quick succession. At first they change hue to a grayer, more modest version of their summer plumage, as the colors around them also grow tired and faded. Then, later on, as the tundra succumbs once again to the grip of the long Arctic winter, the changes are reflected in the patterns of the birds' feathers. White blotches appear on the plumage, which grow and coalesce until, once again, the ptarmigans are barely visible against the snowy landscape.

Most effective camouflage

NAME **willow ptarmigan** *Lagopus lagopus*
LOCATION circumpolar Arctic regions
ATTRIBUTE equipped to hide at any season

Jacanas are a superb example of a family of birds adapted in a unique way to a unique habitat. They would look like any other wading birds if it wasn't for their most obvious feature—their remarkably long, spidery toes. No other birds' toes splay out as far in proportion to their size. On the other hand, no other birds live out their entire lives on the delicately unstable world of floating lily pads.

The mechanics of the matter are simple enough. The longest toes on the largest comb-crested jacanas extend 8 inches (20 cm), and each foot can cover a surface area of about 45 square inches (300 sq. cm). This simply spreads out the bird's weight so that it doesn't sink when it is walking over floating vegetation.

It's an all-or-nothing adaptation. All aspects of their lives, including courtship display and breeding, take place on the jacanas' literal equivalent of thin ice. This places particular emphasis on the nest, which invariably floats on the surface too. Most jacana nests are small piles of vegetation; the male builds several at a time, and when the birds copulate, it is possible that the female monitors how well the nest copes with the weight of two birds before choosing to lay her eggs in a particular one.

Longest toes

NAME **comb-crested jacana** *Irediparra gallinacea*
LOCATION Australia
ATTRIBUTE extended toes

Strongest claws

NAME **harpy eagle** *Harpia harpyja*
LOCATION Central and South America
ATTRIBUTE very strong and long hind claws

The magnificent harpy eagle is the largest predatory bird in the world. Some vultures are larger, but they are mainly carrion eaters, taking only occasional live prey. This eagle is a professional killer, terrorizing virtually every medium-sized mammal and large bird that crosses its path.

The vital statistics of this great hunter are awesome. It is more than 3 feet (1 m) in length, with a wingspan of nearly 6½ feet (2 m) and can weigh 9–20 pounds (4–9 kg), and the females are much larger and heavier than the males. Despite its size, this eagle is surprisingly agile, able to negotiate its way through thick forest and to drop rapidly onto prey from a perch. Chillingly, its hind claw is nearly 3 inches (7 cm) long and as sharp as a knife.

Not surprisingly, the harpy eagle holds the avian record for seizing and carrying off the largest prey item—a red howler monkey (*Alouatta sara*) weighing 15 pounds (6.8 kg). Apart from monkeys, harpy eagles take a variety of different creatures as prey, of which sloths are a particular favorite. The latter may constitute a significant part of the harpy's diet; it is thought that their habit of sunning themselves in the treetops in the early morning makes them especially vulnerable to predation. Harpies also catch domestic pigs and goats, young peccaries, plus armadillos, porcupines, foxes and macaws. They can also deal with snakes approaching 2 inches (5 cm) in diameter by slicing them in two.

Up until now there have been no records of harpy eagles attacking humans. But if you do find yourself alone in the Amazon rain forest, make sure you keep a good lookout up above!

Biggest mouth

NAME **tawny frogmouth** *Podargus strigoides*
LOCATION Australia
ATTRIBUTE wide gape

There are a few bird families vying for the title of having the biggest mouth for size of bird, but perhaps the frogmouths of Australasia and the Far East have the best claim, not least because of their name and the unusual way they are thought to use their gape. As the photograph demonstrates, the enormous head gives the frogmouth a most peculiar, unbalanced look, with a huge front end and a truncated posterior.

The tawny frogmouth has a gape about 2 inches (5 cm) wide and can open its mouth to about the same depth. This allows plenty of room to accommodate a variety of animal prey, including the largest insects and spiders and also, occasionally, frogs, lizards, small mammals and even birds, which are usually caught on the ground after a brief sally from an elevated perch. Most are swallowed whole, but pesky prey that tries to fight back is shaken to death or struck against a hard surface to break its bones and any further resistance. There is no easy escape from the frogmouth's heavy, sharp-sided bill.

Some observations, albeit not yet confirmed, suggest that frogmouths might use their mouths in another, less conventional way. The tawny frogmouth is sometimes observed perching with its mouth wide open for minutes on end, bar the occasional snap shut. It is thought that the mouth may exude a smelly saliva that acts as bait to attract insects. If so, this would be unique among birds.

Smelliest bird

NAME **crested auklet** *Aethia cristatella*
LOCATION Bering Sea
ATTRIBUTE strong body odor

Distinctive though the crested auklet undoubtedly is to look at, this is one species that bird-watchers can identify with their eyes closed—as long as the wind is blowing in their direction. For the crested auklet, together with its close relative the whiskered auklet (*Aethia pygmaea*), is just about the smelliest bird in the world.

Colonies of these birds breed on islands in the Bering Sea. Visitors approaching by boat have been known to detect an auklet colony from as far away as 6 miles (10 km)—all because of a peculiar and very distinctive tangerinelike odor that is unique to both species.

The most interesting question about the smell is what is it for? The answer is that no one really knows, although the complex biological processes involved in developing such a unique odor must surely attest to its usefulness. The most plausible explanation is that the birds might use the smell to detect their colonies when they are returning from fishing at sea. Or perhaps they use the smell to home in on the colony where they were born, in the same way that salmon use scent to return to their natal stream.

Oddest bill

NAME **black skimmer** *Rynchops niger*
LOCATION the Americas
ATTRIBUTE peculiarly shaped bill

It's a perfect marriage of form and function. A black skimmer flies over a shallow lagoon, slicing its lower mandible through the water in an effort to detect fish by touch while lifting its upper mandible up out of the way. No other bird feeds regularly in this way, and no other bird has a bill anything like a skimmer's.

The skimmer's bill is long and large, but its most distinctive feature is that the lower mandible is longer than the upper by ½–1¼ inches (1–3 cm), creating a weird appearance, unique in a bird. But this discrepancy in jaw length is actually a by-product of the bird's feeding method, rather than at the heart of it. The act of skimming over the water surface induces a great deal of wear and tear, and from time to time the tip of the lower mandible breaks when it hits an object. Thus, the horny distal part of the bill, the rhamphotheca, grows continually, like a fingernail, at a greater rate than that of the upper mandible, producing the imbalance.

Functionally, the most extraordinary part of the bill is the lower mandible's bladelike tip, which literally slices through the water with minimal resistance. It is so narrow that it cuts through the water with little wake, and it is this part that comes in contact with fish close to the surface; farther up the bill widens considerably. Once a strike has been made, the fish is grabbed in both mandibles by a lightning-quick bend of the neck, with the head frequently facing backward as the fish is caught.

There's nothing subtle about the claim to fame of the sword-billed hummingbird, as the photograph opposite shows. The remarkably extended bill is almost as long as the bird itself. The male averages 5½ inches (14 cm), with his bill stretching to 4 inches (10 cm); the female is 5⅛ inches (13 cm) long, and her bill is a staggering 4½–4¾ inches (11–12 cm) in length. No other bird in the world comes close to matching this species in terms of ratio of bill to overall length.

It is equally obvious why the sword-bill should be so endowed. In common with dozens of other hummingbirds, its bill is perfectly shaped to match the particular bloom from which it drinks. As plants and birds have coevolved, ever more extreme blooms have provided niches for ever more extreme bill shapes; the bird beats the competition, the plant assures its pollination. As for the sword-billed hummingbird, it specializes in plants with long, hanging blooms, such as *Daturas*, passion flowers and fuchsias. These may not grow in the same profusion as other plants, but the nectar supply that each provides is rich and generous.

Of course, being so hyper-adapted causes some difficulties for this hummingbird. It must always perch and fly with its bill held up at the steep angle shown, otherwise it would become overbalanced. And preening can be a problem too: with the bill permanently out of action, the bird has to perform feather care with its feet!

Longest bill

NAME **sword-billed hummingbird** *Ensifera ensifera*
LOCATION Andes
ATTRIBUTE longest bill of any bird in relation to body length

Best energy saver

NAME **greater roadrunner**
Geococcyx californianus
LOCATION southern United States and Mexico
ATTRIBUTE heat-absorbing skin on its back

Just in case you thought that the roadrunner was merely a fictional cartoon character, the photograph opposite should convince you that it is alive and well and living outside Hollywood. It holds the record for being the fastest-running bird that can fly, reaching an impressive 15 mph (24 km/h) when pressed, although sadly it does not go "beep, beep" as it does so.

In fact, the real roadrunner is a member of the cuckoo family and is a marvelous example of a creature finely attuned to the desert environment, having a number of adaptations enabling it to cope with the notoriously variable temperatures of the desert, where the mercury soars by day and plunges by night. To meet these demands the roadrunner has resorted to acting like a reptile. Although it is warm blooded, when the temperature falls after dark the roadrunner begins to turn down its own thermostat, dropping its metabolic rate and its body temperature by a few degrees, and cutting down on energy costs. It plunges into a slightly torpid state.

In the morning it is important for the roadrunner to warm up quickly, and this it does in a highly unusual way—by using its own solar panel. On its back it has a patch of feathered skin heavily pigmented with melanin, a chemical that absorbs sunlight. The bird fluffs its feathers and the light efficiently penetrates to the skin, passing the precious heat onto the blood vessels and from there through the body. After half an hour or so of basking the bird is ready to resume running, having saved 50 percent of the energy it would have needed to rouse itself without the benefit of its very own solar-powered heating.

Most poisonous bird

NAME **hooded pitohui** *Pitohui dichrous*
LOCATION New Guinea
ATTRIBUTE poisonous feathers

In the summer of 1989 a research biologist called Jack Dumbacher was studying birds of paradise in New Guinea. The project involved catching and banding the target birds and, at the same time, releasing the many other species caught in the mist nets alongside them. Among these birds were hooded pitohuis, common birds of the forest.

One day Dumbacher was attempting to release yet another collateral pitohui when the bird ungratefully pecked and scratched him. The field-worker put his bleeding fingers in his mouth and immediately felt a curious numbing sensation. He recognized the feeling as being caused by a toxin, but it was not until he licked one of the bird's feathers that he realized that these, and not some local plant, were the source of the poison. He had stumbled upon the very first recorded instance of toxicity in a bird's plumage. Despite the fact that pitohuis had been known to science for hundreds of years, with dozens of specimens in museums around the world, nobody had noticed the phenomenon before.

The toxin in the pitohui's plumage is the same neurotoxin found on the skin of poisonous frogs in South America. Although there is enough poison on a pitohui's skin and each of its feathers to kill several laboratory mice, the levels are still much lower than on the frogs, and it is unlikely that a predator attacking a pitohui would be killed. Nevertheless, the effects would be sufficiently unpleasant at first bite to ensure that predators would give the bird a wide berth in future. It is also possible that the toxins help to keep parasites off the pitohui's skin and feathers.

Snuggest underwear

It can be cold up there on the windblown steppes of central Asia, where the Pallas's sandgrouse lives a mysterious life under the cover of its cryptic plumage and even more secretive habits. By contrast the temperature can be ferociously hot in the middle of the day, to the point where everything shuts down and retreats to the shade or a state of semi-torpor. The extremes of this climate represent the greatest challenge to the local fauna. It is thus that the Pallas's sandgrouse, and other sandgrouse that live in the arid regions of most of the Old World, have developed some highly unusual adaptations to cope with the great fluctuations in temperature.

The most interesting of these adaptations—one that is unique among birds—is something akin to thermal underwear. While most species of bird have down feathers arranged in tracts below the rest of their plumage, with gaps between each tract, sandgrouse have an unbroken layer of black furry down. Furthermore, the contour feathers that form the exterior of the bird are particularly downy—or "plumulaceous"—at the base. Nothing creeps through this protective layer, not even on the legs and toes, which are feathered rather than bare.

The thermal covering is ideal for temperature regulation, but it is not the only item in the sandgrouse's armory. These birds also have a greater capacity for cooling down than most other birds, achieving it by water loss through evaporation from the skin.

NAME **Pallas's sandgrouse**
Syrrhaptes paradoxus
LOCATION central Asia
ATTRIBUTE allover cover of down

It only takes a glance at the photograph to appreciate the self-evident hideousness of the marabou stork. Surely only other marabous could possibly find this bird attractive? Nonetheless, if we can take a step back and be a little more scientific, surely it should be possible to find some redeeming features?

The trouble is, as we delve deeper, we find more to make us recoil. The marabou stork is a bird of death, a scavenger, that spends a good deal of its time around corpses, sometimes those of humans, picking off little bits of flesh at the edge of the bone-stripping scrum. When it is not mixing with hyenas, jackals and vultures at the scene of death, it visits human garbage dumps instead, where it is best not to elaborate on its diet. In a further blow to public relations it regularly kills pretty flamingos in front of tourists, and it also has the insalubrious habit of defecating on its legs to keep cool. All in all, marabou storks do nothing to endear themselves to us.

Of course it makes sense for a scavenger to have a bare head and neck. All that blood and body fluid would cake the plumage and cause problems. The enormous bill is lighter and more sensitive than it looks, and the ugly sac hanging down from the throat is useful for temperature control and communication between storks. But few birds anywhere in the world have such a combination of features that we humans find unsightly. Even for such a subjective assessment, few of us are likely to protest at this species being declared the ugliest bird in the world.

Ugliest looks

NAME **marabou stork** *Leptoptilos crumeniferus*
LOCATION sub-Saharan Africa
ATTRIBUTE being shockingly ugly

Lightest bones

NAME **great frigate bird** *Fregata minor*
LOCATION tropical seas
ATTRIBUTE lightest skeleton of any bird

Frigate birds are among the most recognizable of all species. Their dark plumage, long wingspan, streamlined body and long, forked tail make them both unmistakable and somewhat menacing, especially when they sail overhead like giant swallows and peer down with their long, hooked bills seemingly at the ready. The male great frigate bird shown here also has his red throat sac inflated, the sac being used as part of the courtship display.

Although very much seabirds, frigates cannot swim properly and hardly ever enter the water. Their feet are weak and barely webbed, while their plumage isn't waterproof. Yet they are often found hundreds of miles out over the ocean, where they will make their living by snatching fish or squid from the surface of the water with a dexterous downward swoop and snap of the bill. Frigate birds are also expert in robbing food from other birds, including their own species. They do this by aerial pursuit, homing in on an adult booby or tern returning to its nest and tailgating it in midair until, in distress, the victim drops its quarry. Some individuals engage in this particular form of abuse for weeks on end.

The sight of a frigate bird harassing another seabird demonstrates its peerless expertise in the air, a result of a number of remarkable adaptations. Frigate birds have the lightest skeletons of any birds—it accounts for only 5 percent of their total weight. The bones are partly hollow and filled with air, and, amazingly, their total weight adds up to less than the weight of the bird's feathers. A frigate bird's enormous wings mean that it also has the lightest wing loading (the ratio of wing area to overall weight) of any bird, while its angular shape confers upon it the ability to twist and turn with supreme agility.

This photograph is unusual. Something has obviously disturbed this kori bustard because it has broken the habit of a lifetime and taken to the air. It would much prefer to be strutting across the African savanna like a wannabe ostrich.

Nonetheless, bulky and ponderous though it looks, the kori bustard can undoubtedly fly, and as such it is possibly the heaviest flying bird in the world. An adult male has been recorded at 42 pounds (19 kg), the weight of a small child. However, as is typical for such records, there are other claimants, in particular the closely related great bustard (*Otis tarda*), for which 47 pounds (21 kg) has been claimed, and the mute swan (*Cygnus olor*) at an incredible 53 pounds (24 kg). In the end, all these birds are on the upper limit for what is actually possible, given their physiology. It has been demonstrated, for example, that for sustained flight birds struggle if their weight exceeds 26 pounds (12 kg).

Kori bustards probably never fly very far in any case. They are fast runners and can usually escape from predators that way. They are also wary and cryptically camouflaged, so they are less likely to be seen in the first place.

Some evidence suggests that kori bustards do travel a little, moving small distances after breeding and sometimes changing their habitat to a more wooded, enclosed one. But whether they do so by walking or flying is unclear.

Heaviest flier

NAME **kori bustard** *Ardeotis kori*
LOCATION east and southern Africa
ATTRIBUTE flying despite its great weight

Worst flier

NAME **spotted nothura** (a type of tinamou)
Nothura maculosa
LOCATION South America
ATTRIBUTE flying very badly!

Birds are renowned for their powers of flight. The grace of an albatross is proverbial, the maneuverability of a hummingbird breathtaking. However, the tinamous, of all birds, demonstrate just how difficult flight can be, in an almost comical statement of incompetence.

Tinamous, such as this spotted nothura, spend most of their lives on the ground; they only need to fly when under extreme pressure, for example, from ground predators. They are heavy, muscular birds with big feet, but their wings and tail—aerodynamic essentials—are decidedly small for the size of bird. Often when tinamous make an emergency takeoff their control lets them down, and they find themselves crashing into objects, such as trees or rocks, often killing themselves. Landing is also no better than a controlled crash, and they regularly touch down at a run, legs a blur.

Once they do get airborne, tinamous are not built for the long haul either. Far from it—1,640 feet (500 m) is about as far as they can go. The problem is their circulation. They have unusually narrow blood vessels, as well as the smallest heart, relative to their size, of any bird. Their lungs aren't much better.

Lest you should think, though, that tinamous are poorly adapted for daily living, you should be aware that they are a highly successful family, with 46 living species, and their range covers all of South America north to Mexico. They aren't a failure. It's just that their particular conquest of the world starts, and ends, on the ground.

Best submarine

NAME **darter** *Anhinga melanogaster*
LOCATION warmer regions of Africa, Asia and Australasia
ATTRIBUTE plumage that gets soaked, enabling the bird to sink

This picture of a darter, half submerged in the water with only its head and neck showing, is typical of the species. Of all the birds in the world, the darter, along with its American counterpart the anhinga (*Anhinga anhinga*), has the most control of its own buoyancy. It acts like a mini-submarine, of which the head and neck are the periscope.

The idea of a water bird primarily adapted to sink might seem slightly absurd in a world where water proverbially flows off a duck's back, but the darter's niche is unique. Instead of chasing after prey, it is actually an underwater stealth hunter that lies in wait for its prey to come within striking distance. Instead of swimming around, it lurks in the shadows, often splaying out its wings below the water to create its own shade. In many ways it hunts like an underwater heron and, in common with those birds, it has special cervical vertebrae that enable its neck to be retracted into an S shape and then thrust out like a harpoon. In this way fish are often impaled on the darter's mandibles.

In order to sink effectively darters have some unusual body modifications. While most birds have light, hollowed-out bones, those of the darter are denser and heavier than usual. Meanwhile, the plumage is specially adapted to become completely saturated when submerged. The microscopic structure of the feathers allows water to fit into the spaces inside them, making the darter's feathers three times more permeable than those of a cormorant, for example. Thus, when a darter enters the water its buoyancy is almost neutral, and it can have complete control over its movements without having to worry about its body being forced to the surface.

Most flexible mouth

NAME **Eurasian nightjar** *Caprimulgus europaeus*
LOCATION Eurasia
ATTRIBUTE jaws that move sideways as well as up and down

A nightjar sits on a branch, roosting, concealed by its cryptic plumage and its lack of movement. A nocturnal hunter, its days are spent hiding away; only at night does it come alive.

Nightjars utilize a food supply that is abundant but hard to get at—insects that fly at night. By day hundreds of species of birds catch insects in quick, airborne sallies; by night they sleep and leave the plenty untapped. Supremely well adapted for hunting in the dark, nightjars take full advantage of this.

Skimming through the air to catch insects presents two main problems: seeing the prey and snatching it. You might not have thought of the first as a problem because it is easy to imagine a bird trawling through the air with mouth wide open, catching food randomly. But nightjars don't find this profitable. Instead, with eyes that are adapted to detect the faintest light and to see contrast, the birds target their prey and home in on it. They often feed preferentially at dawn, dusk and by moonlight.

The second problem, obtaining the prey, is solved by modifications to the mouth. The nightjar's bill is small, but the gape is relatively enormous. The palate and the bristles on the side of the mouth act as touch detectors, enabling the bird to sense when it has caught something. And, more than anything, the nightjar has the most astonishingly flexible jaw; not only can its jaw open very wide, but it has special modifications that enable it to move to the side, which greatly extends the reach of the mouth.

Oddest skin

NAME **southern screamer** *Chauna torquata*
LOCATION South America
ATTRIBUTE skin that crackles when touched

The peculiar bird in the picture is a member of a somewhat obscure family of South American birds known as the screamers. Despite their tall stance and hooked bills, screamers are thought to be closely related to ducks and geese, although they have a few anatomical oddities that are very much their own. One of these is their skin.

If you were ever to reach out and touch a screamer (admittedly, not very likely), you would hear a distinct crackling sound. This is caused by a complicated system of small air sacs separating the outer skin from the rest of the body, which contract to make the sound, like the bursting of hundreds of minute balloons. It is thought that these air sacs play a part in insulating the bird.

Another curious feature of screamers is that their bones are more highly pneumatized than those of any other bird—in other words, their internal structure is more hollowed out, making the bones exceptionally light. Oddly, the screamer's rib cage also lacks the special strengthening known as the "uncinate process" that is present on all other bird bones. All in all, it seems, the sheer weirdness of screamers is more than skin deep.

Longest tongue

NAME **black woodpecker** *Dryocopus martius*
LOCATION Eurasia
ATTRIBUTE unusually long tongue

The tongue of this black woodpecker looks almost snakelike as the bird prepares to lap up some of its favorite food—ants. Black woodpeckers may spend hours at an ant nest if undisturbed, working away mercilessly at the fast-depleting colony.

Woodpeckers are so well known for their highly modified bills, skulls and gripping feet that another of their anatomical peculiarities, their long tongue, is often overlooked. It actually protrudes beyond the tip of their bill, further than that of any other bird. In the case of the black woodpecker, it can amount to a distance of 2¼ inches (5–5.5 cm), while the European green woodpecker (*Picus viridis*), another ant guzzler, has a tongue that protrudes up to 4 inches (10 cm)—on a bird only 13 inches (33 cm) long.

It isn't only the tongue's length that is unusual. A woodpecker's tongue is capable of independent movement at the tip, which makes it a formidable weapon for lapping up prey hidden away in holes. The sublingual gland secretes a sticky fluid that causes food to adhere to the tongue's surface. Meanwhile the tongue is also specially hardened to enable a woodpecker to impale soft-bodied creatures, such as grubs or caterpillars. In the case of the black woodpecker, the tip is further armed with clusters of three to five backward-pointing barbs or hooks, which cling onto the bird's prey like harpoons stuck into the back of an unfortunate whale. This allows the woodpecker to extract food easily, no matter how deeply the food is stuck in a crevice.

A woodpecker's tongue is controlled by a complex system of bones and muscles known as the "hyoid apparatus," which takes the form of two "horns" that sweep behind the tongue and wrap all the way around the skull. Those ants don't stand a chance!

Most variable feathers

NAME **raggiana bird of paradise**
Paradisaea raggiana
LOCATION New Guinea
ATTRIBUTE multiple designs in feather structure

A raggiana bird of paradise perches on its branch in the mid-levels of a New Guinea forest, displayed like an overdressed model posing for a fashion shoot. The extraordinarily silky red plumes draped over the bird's back are not, as it would seem, tail feathers but actually originate from under the wings, on the flanks.

Birds of paradise are famous for their extraordinary plumes. Indeed, their opulence has earned them a curious mix of awe and exploitation in their native lands. In New Guinea some local peoples still wear bird of paradise feathers in their everyday dress, while the more showy costumes are used ceremonially; at the same time, tribal dances incorporate themes from the routines of the displaying birds. Feathers have been used for trade in the region for centuries, although often to the detriment of the birds that bear them.

The raggiana is just one species among 40 or so distributed across New Guinea and Australia, and, as a family, the birds of paradise exhibit the most extravagant variety in feather structure of any bird family in the world. An astonishing range of shapes and colors crops up in the plumage across the different species. The tail feathers, of which there are 12, are particularly variable, from those pared down to nothing but wiry shafts in the twelve-wired bird of paradise (*Seleucidis melanoleuca*) to enormous 3-foot (1 m) long sabers in the black sicklebill (*Epimachus fastuosus*).

Equally bizarre feathers spring from the head. In the parotias (*Parotia* spp.) there are six occipital plumes that are entirely bare except for a blob at the end, while in the astonishing King of Saxony bird of paradise (*Pteridophora alberti*) a single feather sprouting from behind each eye grows to twice the length of the rest of the bird. It is waxy blue and only intermittently veined, making it look like flags on a string. Other amazing feathers spring from the crest, back, breast and wings, depending on the species.

Heaviest testes

NAME **alpine accentor** *Prunella collaris*
LOCATION mountains of Eurasia
ATTRIBUTE heaviest testes in relation to body size

Handsome but hardly exceptional looking, the alpine accentor holds what must be one of the oddest of all bird records. Males in the breeding season possess enormous testes, which constitute an unprecedented 8 percent of their entire body weight. This statistic, possibly to the delight of arch-feminists everywhere, is considerably more than the weight of the bird's brain.

But why should this unremarkable-looking bird have such large testes? Of course, the ordinariness of the bird's appearance is merely subjective; what is exceptional is the alpine accentor's breeding system. Up in the high mountains of Europe and Asia, the patchy nature of the food supply has rendered being territorial somewhat fruitless. Instead, birds live in groups of mixed membership, within which males and females tend to have multiple partners—the precursor to sexual competition.

The interrelationships in alpine accentors are complex. However, broadly speaking, within each group every male competes with every other male for access to females (and thus paternity). Every female then competes with her rivals for paternal care. The latter is achieved by each female, ensuring that she copulates with as many males as possible. This, in turn, ensures that it will be in each male's interest to feed her offspring.

On the other hand, the battle for paternity is less straightforward. Essentially, a male can copulate with each female, but that doesn't, in itself, guarantee paternity. To secure this role, a male must ensure both that he mates regularly with each female, and that, when he does, he releases enough sperm to swamp the contributions of his rivals. This, therefore, explains the need to develop such large testes.

It could be argued that some bird or another was bound to have the longest penis in relation to body size. Indeed, on reading this you may be wondering how well qualified the lake duck really is for an entry in this book.

Well, this record needs to be put into context. Most birds don't have a penis at all. Both males and females have an external reproductive opening, the cloaca, and copulation normally requires just minimum cloacal contact for sperm exchange to occur. A few families possess an external male organ, including the large flightless birds such as ostriches, in addition to tinamous, ducks, storks, flamingos, game birds, thick-knees and, bizarrely, a small number of weaverbirds from Africa. Why some birds do and some birds don't have a penis would make a book in itself.

On the whole one might expect the largest of birds to have the largest penis and, indeed, that of the ostrich, a bird nearly 7 feet (2 m) tall, may be up to 8 inches (20 cm) long. But, incredibly, even some ostrich's vital statistics are beaten by a small diving duck found on the South American lakes.

The sheer size of the lake duck's penis is mind-boggling. The bird itself is only 18 inches (46 cm) long at most, yet its penis ranges in length from 7½ to 9¾ inches (19–24.5 cm)—half the entire length of the body. The surface is covered in dense spines, and the penis is spiral shaped, like a corkscrew, so that it can be retracted after use.

Why on earth should this bird be so extravagantly endowed? Well, the lake duck is sometimes promiscuous, and it is thought that the size of its penis plays a part in what is clearly unsubtle sexual selection. Birds also sometimes practice forced copulations, and it is believed that the penis may displace the sperm of rival males. Such behavior, however, applies to a number of species with nothing but cloacas, so the penile development of the lake duck is a remarkable trait indeed.

Longest penis

NAME **lake duck** *Oxyura vittata*
LOCATION southern South America
ATTRIBUTE largest penis for the size of bird

Sexiest tail

NAME **barn swallow** *Hirundo rustica*
LOCATION worldwide
ATTRIBUTE telltale tail

There are plenty of birds in the world that attempt to impress potential mates with their tails. The peacock, for example, demonstrates a good example of rear-end sexual selection. But there can't be too many species in which we, as people, can tell at a glance what an individual's prospects for the breeding season will be. With the barn swallow, however, we can.

Studies in Europe have shown that the best males have the longest tails and also the most symmetrical ones, so the example here, sadly, is not a top-quality bird. Males with long, symmetrical tails are more resistant to disease and parasites than those with shorter, uneven tails, and their whole breeding season passes with serene efficiency. The superior individuals acquire mates more quickly than the rest, their mates are of higher quality than those available to nonsymmetricals, and they bring up more young.

The inferior males, on the other hand, have to work harder for everything. Notably, if they do succeed in obtaining a mate, they will invest much more effort in feeding the young than the long-tailed males will. Unfortunately, however, despite their best efforts, their mates often slip away in quiet moments to copulate with any high-quality males on hand, so even if the inferior males do rear clutches, not all the young will be theirs.

Why should the tail be such a symbol of superiority? It seems that there comes a point when a long tail is an encumbrance to birds, especially an aerial species, such as the barn swallow. Males that grow them, despite this difficulty, are showing that they have what it takes to survive against such odds. It is a powerful statement of their health and fitness.

Biggest belly

NAME **hoatzin** *Opisthocomus hoazin*
LOCATION northern South America
ATTRIBUTE vastly expanded foregut

A hoatzin tends a chick on its large stick nest beside a quiet backwater in Guyana. Despite the homely scene, the hoatzin still looks like the decidedly peculiar bird that it is, with its wild-eyed expression, wispy crest and bulging belly. The latter is a particularly prominent and unflattering feature, obvious at all times. When resting, hoatzins don't normally perch but instead sit on their sternum, tending to look almost comically fed up and unmotivated. Their bellies can be so big and heavy that, when bending down from a low branch to take a sip of water, these birds have been known to topple over and splash calamitously in.

The large belly is the result of the hoatzin's unusual and restrictive diet. It is one of the few birds to specialize in eating leaves, which can constitute up to 82 percent of its entire intake. Leaves are notoriously difficult to digest, and hoatzins, in common with many herbivores, rely on microbes to do the work for them. Microbes work best in a capacious gut where the flow of matter is not too fast, and both the acidity and temperature remain constant. This partly explains the large size of the belly. However, in a break from all other birds, hoatzins provide these conditions in their foregut, instead of further back in the alimentary canal. Digestion takes place in the crop and esophagus, which is more akin to the system in sheep, cows and kangaroos than the cecum-based digestion (in the large intestine) seen in birds, such as grouse.

Almost everything about this system seems awkward for the hoatzin. It cannot fly far, it cannot swim and it has great difficulty moving through the branches of the waterside plants where it lives, frequently damaging its feathers as it crashes around. Still, leaves are never far away and, in the warm, stable conditions of tropical waterways, this bird thrives as well as any of its more energetic neighbors.

The portrait opposite shows a Fischer's turaco, which you will doubtless admire as a handsome and colorful bird. It would be a travesty if you didn't, because its feathers happen to be suffused with some of the rarest pigments in the entire animal kingdom. Effectively this bird is a model, and it's wearing designer plumage.

The two special pigments are called turacin and turacoverdin, named after the turaco family (Musophagidae)—a small group of 23 species found only in sub-Saharan Africa, of which the Fischer's turaco is one. Both pigments are copper based, and, so far, they have not been found in any animal other than turacos. Indeed, not all turacos have them: a couple of species seem to be lacking them.

Turacin is a red pigment that is mainly found on the wings, although on this Fischer's turaco it can also be seen to adorn the crest and nape. Turacoverdin is a green pigment found throughout the body, and its intensity is related to the habitat of the relevant species: Fischer's turaco, for example, is a forest bird and has it in abundance. Even more interestingly, this is the only green pigment synthesized by any bird. All other green colors, from the wondrous iridescence of hummingbirds to the plainer green of warblers or finches, arise through structural modifications of the feathers that cause light to be refracted unequally.

Young turacos don't manage to acquire full adult colors until they are about a year old. It seems that the required amount of copper upon which the pigments depend takes that long to accumulate from the birds' diet. The turacin found on a single bird can apparently yield about $^{1}/_{3,600}$ ounce (8 mg) of copper—and that's without taking the turacoverdin into account.

Classiest colors

NAME **Fischer's turaco** *Tauraco fischeri*
LOCATION coastal East Africa
ATTRIBUTE astoundingly rare pigments in its plumage

Whitest bird

NAME **ivory gull** *Pagophila eburnea*
LOCATION high Arctic
ATTRIBUTE entirely white plumage

An adult ivory gull flies past, barely visible against the low light of the Arctic sky. Its ethereal plumage is unusual: only a handful of birds in the world are completely white. Most have at least a few darker patches or stains here and there.

It is perhaps not surprising that the ivory gull is an exception, for it spends virtually all its life amid the pack ice of the extreme north, catching fish and feeding on carrion and waste, including dead sea mammals and their feces. It is a relatively small gull, and no doubt the plumage helps to conceal it from predators, especially when it is standing on the snow and ice. However, the color of the underparts also serves an additional purpose. In common with most gulls, ivory gulls take fish, often by dipping down in flight to the water below. Carefully conducted experiments have shown that white plumage conceals avian predators against the sky, not allowing the fish to detect the danger.

Concealment is not always the purpose behind white plumage however. A few other white birds, such as egrets and cockatoos, have white plumage to make themselves more conspicuous, not less. This enables such birds to advertise their presence to other members of the same species, either to gather together or, in the case of egrets, to indicate that a territory is filled. In fact, ivory gulls sometimes gather into flocks, and on such occasions it is even possible that their multipurpose whiteness helps them draw together.

Most variation in a species

NAME **ruff** *Philomachus pugnax*
LOCATION northern Eurasia
ATTRIBUTE greatest variation among individual males

These two ruffs might look as though they are sharing a secret, but in fact they are fierce rivals. You could say they are diametrically opposed, in both motivation and in the color of their plumage.

The ruff has an unusual breeding system in which males and females meet only for copulation. The meeting places are known as leks, and most leks are attended by 10 or so males who display directly against each other in full sight of everyone. When a female visits, all the males ruffle their magnificent head and neck feathers and hope that they, rather than a rival, will be chosen as the female's mate.

The female has a lot of choice. Male ruffs don't just differ in the strength and vigor of their display; they are all individually different to look at to a degree unique among birds. The two birds here—one with white ruffs and the other with black—are but two examples of this variation. There are also brown-ruffed males and reddish males, with no two birds looking the same. From April until June each year male ruffs are the most variable in appearance of any bird in the world.

Intriguingly, the specific color of an individual ruff is related to its behavior. The dark ruffs, including those with black or brown coloration, are known as "independents"; they hold season-long territories at a specific arena and spend their time attempting to lure the visiting females. The white ruffs, on the other hand, are known as "satellites." They don't hold territories at all but move from lek to lek, and they tend to "cheat" opportunistically, copulating with females when, for example, an independent's back is turned, distracted by a rival or another female.

Why should the independents tolerate the satellites? The most plausible theory holds that the satellites' white plumage makes the lek more conspicuous and attracts more visiting females.

A great egret raises its plumes in a courtship display. Each bird has 30–50 elongated shoulder feathers, each about 20 inches (50 cm) long, allowing it to transform itself into a poor man's peacock.

Best grooming aids

NAME **great egret** *Ardea alba*
LOCATION almost worldwide
ATTRIBUTE powder and a comb for grooming

The heron family has a bit of a thing about plumage care, its members spending inordinate amounts of time preening and scratching. This is partly to do with their waterside habitat, in which the feathers soon get wet and soiled. Their main prey, fish, are also oily and scaly, adding to the stress on the plumage.

Herons and egrets have two unusual adaptations to enable them to cope with these issues. The first is powder, which is manufactured on the breast and rump. It begins life as down, the small fluffy feathers on the skin of birds, which quickly disintegrates as it grows. The powdery remains can be preened onto the plumage, where they act as a water repellent and perhaps also as a coagulant for oil and water. The second adaptation is on the middle toe of each foot, where the claw is pectinated—that is, it is fitted with small projections like the teeth of a comb. These help with specialized, precision preening.

Although not strictly equivalent, it is difficult not to imagine the egret using a comb and talcum powder, just as we do. But if you wanted to look as good as this egret you'd clearly need to put in the hours!

Most useful esophagus

NAME **greater prairie chicken** *Tympanuchus cupido*
LOCATION central plains of North America
ATTRIBUTE inflatable esophagus

A male greater prairie chicken inflates the strange orange-yellow sacs on his neck to create an impression on a female during his display. The sacs are offshoots of the esophagus, which for us is merely the conduit for food between the mouth and the stomach. The esophagus serves that role in the prairie chicken, too, but it has been modified to carry out two other tasks as well.

The first task is to look good in order to attract a mate. For the most part, the air sacs are left empty, hidden beneath the prairie chicken's cryptically colored plumage. In early mornings between March and May, however, males gather together at communal display grounds, where they undergo a complex regime of stereotyped actions and postures, including the one seen here. Females visit these gatherings, or leks, to seek out genetic material for their eggs, and they naturally take notice of the ballooning air sacs.

To inflate his sacs the male closes off his nostrils with his tongue, shuts his bill and forces air from his trachea into his esophagus and, thus, the sacs. At the same time the blood supply to the skin of the sacs increases, intensifying the color pigmentation and, hopefully, making the bearer irresistible to the opposite sex.

The esophagus also enables the prairie chicken to make a low, moaning boom—another way to impress female visitors. Vocal in origin, the boom is merely resonated in the sacs to amplify the sound, making it carry far across the prairies. Hence, without some peculiar modifications to a simple organ, the prairie chicken would be bereft of ways in which to make itself a winner with the females.

Extreme

Ability

Fastest swimmer · Most vicious kicker · Longest time aloft · Biggest plunge dive · Longest water dash · Fastest wing beat · Best flock coordination · Longest nonstop journey · Highest migration · Fastest mover · Most elegant dancer · Deepest dive · Fullest bill · Strongest stomach · Choosiest eater · Most patient feeder · Longest fast · Biggest binge eater · Greediest berry eater · Best mimic · Noisiest call · Most tireless singer · Sharpest hearing · Keenest eyesight · Craftiest builder · Highest forager · Biggest communal nest · Most prolific breeder · Strangest way to cool off · Best architect · Smartest transporter · Best thermal engineer · Longest life · Most remarkable immunity

Fastest swimmer

NAME **gentoo penguin** *Pygoscelis papua*
LOCATION Antarctic Peninsula and subantarctic islands
ABILITY moving through water faster than any other bird

At first sight these gentoo penguins would appear to be running on the water rather than swimming through it, but the camera has captured them in the process of "porpoising." It's a type of locomotion that, among birds, is exclusive to penguins. It involves intermittently leaping out of the water and splashing in again, so that the birds alternate mediums as they propel rapidly forward.

Nobody is sure why penguins use porpoising. It might serve to confuse marine predators, in the same way that fish momentarily leap out of the water when they are pursued. More likely it could help the penguins breathe while swimming, or it could increase a bird's overall speed as it cuts periodically through the lighter medium of air. Another possible benefit is that it could regularly replenish the layer of air trapped between the skin and feathers, which tends to be lost in underwater swimming, that helps to keep the bird warm.

Porpoising is one of three modes of swimming utilized by penguins. Sometimes the birds simply idle on the surface, with heads above water, not fully flapping their flippers and not generating any significant pace—a little like our method of "treading water." During underwater dives, however, they employ a flying motion with their flippers—analogous to an aerial bird flapping its wings. During a dive the feet and tail are used only as rudders.

Whether flying underwater or porpoising, penguins generate far more pace through the water than any other bird. Speeds of between 3 and 6 mph (5 and 10 km/h) have been recorded, with 9 mph (14 km/h) claimed for the largest species, the emperor penguin (*Aptenodytes forsteri*). However, because not many porpoising penguins have been followed, it is more than likely that birds in a hurry could exceed the speeds that have been quoted here.

Most vicious kicker

NAME **secretary bird** *Sagittarius serpentarius*
LOCATION sub-Saharan Africa
ABILITY kicking its prey to death

A secretary bird wanders through the grasslands of East Africa on the prowl for food. Its large, sharp, fiercely curved bill indicates its predatory nature well enough, but everything else about this bird is unusual. Its curiously pink legs are three times longer than those of a typical raptor of similar size, making the secretary bird a tall species, some 4 feet (1.2 m) when fully upright. The plumage is also distinctly odd. The multi-plumed crest makes the secretary bird look like an aging hippie, and its spindly legs have dark plumage only down to the "knee" joint (actually the ankle), as if that same hippie was donning leggings for an incongruous attempt to get fit.

Many animals, however, even up to the size of small antelopes, flee when a secretary bird approaches, because to them it is a fearsome instead of a comical sight. This bird exacts a high toll upon the myriad animals of the bush, everything from grasshoppers to large snakes. And it does so by stamping on them and kicking them.

Being the only bird of prey that confines its killing to the ground, the secretary bird is highly adapted for walking and kicking. Its long legs are armed with heavy scaling to counteract wear and tear and potential bites and scratches from its prey. The three forward-facing toes are long and webbed at the base, providing a wide striking area that packs a powerful punch.

The sheer length of the legs, combined with the balancing actions of the wings, allow this hunter to be exceptionally quick and dexterous, as well as powerful, so that it can use its reach to rain kicks down in swift succession, pulverizing its prey into helplessness. Only when the animal is disabled or comatose does the secretary bird stoop down to apply the coup de grâce with its bill.

Different prey requires various refinements of technique. Invertebrates, such as grasshoppers or beetles, merely need to be stamped on a few times, much as we might try to stamp out a small fire. Small mammals or birds might require a quick strike to the head, while large snakes or hares summon an exhibition of fast and relentless footwork that brings to mind the aggressive human sport of kickboxing.

Longest time aloft

NAME **sooty tern** *Sterna fuscata*
LOCATION tropical oceans
ABILITY flying for over four years before landing

In the air above their breeding colony atop a coral atoll, sooty terns prepare to depart for a foraging expedition. They will soon be out on the open ocean, hundreds of miles from land, searching for small fish and, in particular, squid, which they will bring back for their youngsters. They catch food by a quick snatch from the surface, swooping down and up again, maintaining flight all the while.

Terns are superb in the air, with their angular wings and tails and light bodies. Nevertheless, if you had to select a candidate for a species that holds the record for the longest continuous time in the air, the chances are you wouldn't choose a tern. A swift would better fit the bill, or perhaps an albatross or shearwater. Surely not a tern, a type of bird that is most at home perching on sand, rocks or flotsam?

Yet there is strong evidence that a young sooty tern does not touch the ground from the moment it leaves its breeding colony to the time it returns—some four or five years later—as an adult to breed or search for a mate for the first time. Nobody has ever seen a healthy juvenile bird at rest, whether on a beach, rock, buoy or on any floating object. Juveniles evidently maintain a strictly oceanic existence yet, inconveniently, sooty terns cannot swim: their plumage isn't waterproof.

The thought of keeping aloft for four or more years is mind-boggling to us. How do the birds sleep? Why don't they tire? What about storms? As yet, we have no answers to these questions and can merely marvel at the sooty tern's astonishing lifestyle.

Biggest plunge dive

NAME **northern gannet** *Morus bassanus*
LOCATION Atlantic coasts
ABILITY plunge diving into the sea from a great height

While some seabirds use subtlety and guile in their fish-catching exploits, the same cannot be claimed for the world's three species of gannet. These birds take the plunge, in every sense. Their main method of procuring food is to dive headlong from a considerable height into the sea, allowing their momentum to take them deep below the surface and into the realm of their main prey, schooling fish.

There is some uncertainty as to how high these plunges can be—130 feet (40 m) has been reliably recorded, and 330 feet (100 m) has been claimed—but, let's face it, seafarers and fishermen do have a reputation for stretching the truth a little. What is in no doubt is that this is one of the most daring and spectacular ways that birds have developed to gain their sustenance.

Gannets have several modifications to cope with a life of high diving. Their bodies are streamlined, and on the final approach to water they fold their wings right back behind the tail to reduce drag. The skeletal system is suffused with air sacs to lessen the shock when their bodies strike the water, and, as a practical necessity, their nostrils are closed with flaps of skin upon impact.

The sight of one of these white birds plunging is not just exciting for us to witness but also attracts other gannets in the vicinity. Gannets benefit from fishing en masse, since the dizzying sight of dozens of plunging birds confuses and disorients their prey, slowing the school down and making individual fish easier to catch.

Longest water dash

NAME **western grebe** *Aechmophorus occidentalis*
LOCATION western North America
ABILITY running on water as a courtship display

Two western grebes rush in parallel across the surface of a lake in inland North America. It is one of the most famous bird displays in the world. The coordination of the birds' foot movements, allowing them to keep their bodies upright on the water, together with the elegant shaping of the neck and the slight arching of the wings, confers on these otherwise dumpy fish eaters the grace of ballet dancers.

Grebes are the only birds to perform such long display rushes. Their feet are set far back on the body, and they have no tail to speak of—both peculiarities that help the birds rear out of the water. Of the world's 20 or so species of grebe, the longest rushes are performed by the species shown here and its very close relative the Clark's grebe (*Aechmophorus clarkii*); amazingly, they can skim along the surface for up to 66 feet (20 m), leaving a long splashing wake behind them.

The function of this display is usually the obvious one: to form or to consolidate the pair-bond. Thus, not surprisingly, the performers are generally a male and a female. However, this isn't always the case. Sometimes a male will enlist the help of another male and skim across the surface just to catch the attention of a nearby female. Sometimes, amazingly, male western grebes perform with male Clark's grebes, surely the only example in the avian world of two birds of different species combining to display for the benefit of one.

Fastest wing beat

NAME **blue-tailed hummingbird**
Amazilia cyanura
LOCATION Central America
ABILITY beating its wings 80 times a second in normal hovering flight

Nobody would be surprised to hear that hummingbirds fly with the fastest wing beats known among birds. You only have to watch these jewel-like characters hovering in front of a flower to realize that they are aerodynamically exceptional. They can easily outperform every other bird in terms of maneuverability, being the only birds, for example, that can fly backward. Hummingbirds are also perfectly able to fly upside down if it strikes their fancy.

For a small hummingbird, hovering flight is sustained by a wing-beat rate of about 80 per second, although in the largest species, the giant hummingbird (*Patagona gigas*), this goes down to a mere 10–15 per second, not much faster than a big butterfly. When hummingbirds are really excited, however, such as during courtship routines and chases, their rate increases dramatically; some turbocharged individuals exceed 200 beats per second, roughly equal to that of a bee.

This phenomenal wing-beat rate is aided by some unusual modifications to the wing. For example, the bones of the inner wing are reduced so much that the hummer effectively flies with its second, third and fourth "fingers." Furthermore, the shoulder joint allows for considerable movement in all directions, and the wings are powered by unusually efficient muscles. When the bird is flying the tips of the wings transcribe a figure eight, which means that a small change of angle is all that is needed to adjust the hummer's position in the air.

Hummingbirds are so small that you would not expect them to fly especially fast. But you'd be wrong. With such a phenomenal rate of flapping, they can bomb along in excess of 60 mph (100 km/h).

Best flock coordination

NAME **European starling** *Sturnus vulgaris*
LOCATION much of the Northern Hemisphere
ABILITY aerial maneuvers of enormous flocks

It's one of the great sights in nature. A huge flock of European starlings, thousands strong, meets up in the winter twilight and, using the sky as its stage, wheels across the airspace in a breathtaking display of aerial coordination. These flocks have often been described as resembling smoke, or acting like an amoeba, so eerily do they seem to be functioning as a single organism.

Starlings are not the only birds that perform aerobatics in this way. Waders, especially red knots (*Calidris canutus*), do the same thing before entering their roosts at high tide, and other birds in various parts of the world put on a similar display. On a more prosaic level, it is rare to see any bird, in a flock of any size, collide with its neighbor, so birds are clearly very good at keeping in tight flocks.

But how do they do it? There have been some unusual suggestions along the way—extrasensory perception being one. But recently some researchers have found the answer. It seems that every European starling in a flock manages to stay in coordinated step by keeping its eye on just seven of its neighbors, monitoring and following their movements; the rest is due to fast reactions. These seven birds can be any distance away, but the figure of seven is significant.

Now that this puzzle seems to have been solved, however, several other questions still remain. Why, for instance, should birds gather in such large roosts anyway, when they inevitably attract predators? And why should they put on such a song and a dance just before going to sleep?

Longest nonstop journey

NAME **bar-tailed godwit**
Limosa lapponica
LOCATION tundra from Eurasia to Alaska
ABILITY flying nonstop from the Arctic to New Zealand

A small flock of bar-tailed godwits coasts over the surf. Many astonishing feats have humble beginnings, and it is possible that these godwits may have been snapped setting off on one of the greatest journeys undertaken by any animal.

After breeding in eastern Siberia or western Alaska, bar-tailed godwits fly south to winter in eastern Australia and New Zealand. In itself this is not that remarkable; it's a long way, but a number of birds manage similar distances. But what is mind-boggling is that the godwits have been proven to achieve it in a single transoceanic flight, without any stopovers. That means that they fly a minimum of 6,460 miles (10,400 km) without touching dry land.

Think of this for a moment: over 6,200 miles (10,000 km) in one flight. Until recently, not many airliners could get that far on their fuel load. Bar-tailed godwits are not machines, and their fuel is merely animal fat, formed from a diet of small invertebrates. Furthermore, these birds are only 15–16 inches (37–41 cm) long and weigh no more than 22 ounces (630 g).

The flight is calculated to take about 175 hours, or 7.3 days. Before setting off, the birds eat prodigiously and double their body weight, while at the same time certain internal organs, such as the ovaries or testes, shrink greatly in size. Not surprisingly, such an astonishing test of endurance requires intense physiological preparation.

Bar-tailed godwits that have been examined upon arrival, however, have not been found to be emaciated. Far from it—fat reserves indicated that they could fly another 3,100 miles (5,000 km) if necessary. Remarkably, what seems barely credible to us isn't just doable—it's well within their limits.

Bar-headed geese don't have a long migration compared with many birds, but they certainly make their journey in style! These birds breed on the Plateau of Tibet and winter in the lowlands of India. The transfer is only some 450–600 miles (700–1,000 km) in extent, and, with good winds, it can be achieved in less than a day. But what makes the trip rather less than routine is the small matter of the world's highest mountain range, the Himalayas, which lies in the way.

Dozens of migrants crossing Asia avoid the Himalayas, and most steer clear of the highest peaks. The bar-headed goose, however, audaciously takes the most direct route and flies right over them. It has been reliably reported passing Mount Everest at heights of at least 26,400 feet (8,000 m), and it probably exceeds 29,520 feet (9,000 m) at times. No other bird flies so high so regularly.

At such high altitudes humans don't survive for long. Temperatures drop to −58°F (−50°C), and the air pressure is less than one-third of that found at sea level. But bar-headed geese have unusually large hearts, and their lungs are far more efficient than those of people (or any other mammals). Another smart adaptation is that they have at least two, and probably more, types of hemoglobin in their blood. The different types absorb oxygen most efficiently at different atmospheric pressures and are thus most effective at different altitudes.

Highest migration

NAME **bar-headed goose** *Anser indicus*
LOCATION Tibet south to lowlands of India
ABILITY making a regular migratory flight over the Himalayas

Fastest mover

NAME **peregrine falcon** *Falco peregrinus*
LOCATION worldwide
ABILITY traveling faster than any other bird

One of the world's most famous and admired birds, the belligerent peregrine falcon is a powerhouse of aerial destruction. Its bulky but streamlined body allows it to move through the air faster than any other living creature, and its controlled death strike upon its unfortunate prey must rank as one of the most awe-inspiring clashes in the natural world.

What makes the peregrine falcon special is a maneuver known as a "stoop," which is performed at the limits of what any bird can do. A stoop is essentially a dive from a considerable height, using gravity to increase speed as the hunter holds its wings close to its body. It is during this dive that the peregrine has been reliably measured traveling at 112 mph (180 km/h), the fastest recorded intentional speed for an animal (small insects caught up in high winds could be swept along faster). There have been dozens of claims of higher speeds up to more than 186 mph (300 km/h), but for the moment such measurements have not been verified.

There is no doubt that the peregrine falcon is supremely in control of this plunge. When approaching its prey it can make fine adjustments with a flick of the wings. Even when homing in from about a mile (1.6 km) away, it is capable of hitting a moving target, such as a pigeon, straight on. The strike is made by the talons, but the very force of impact is usually enough to break a victim's neck instantly. The death stoop has enabled the peregrine to conquer the skies worldwide. Peregrines occur on all the major continents and are thought to have killed, over time, more than 1,000 species of bird.

Cranes feature heavily in the art of both Japan and China, but it would be hard to imagine a human choreographer producing a display as perfect as the one these birds have created for themselves.

Wherever cranes of any type gather together they "dance." As you can see from the picture, this involves leaping into the air, as well as bowing and running and stretching. At other times a crane might simply stand opposite another bird, bowing its neck and bending its legs, as if curtseying. Material from the ground is often picked up, too, and tossed into the air to add a little extra to the performance. Moreover, it is usually a communal activity; when one bird dances the urge ripples through a flock of cranes, and the leaping transmits from one edge of the group to the other, like a wave.

Why cranes dance is actually surprisingly hard to explain. The birds that dance most frequently are young singles, suggesting that dancing sometimes has a sexual motive. Older birds that are settled into a lifelong pair-bond also dance, probably to ensure that they are behaviorally and physiologically in unison for breeding. But young crane chicks dance in the nest, and adults dance as a displacement activity, so this most elegant of bird displays is as tricky to understand as it is spellbinding to witness.

Most elegant dancer

NAME **red-crowned crane** *Grus japonensis*
LOCATION eastern Asia
ABILITY dancing

Penguins are the ultimate underwater birds. They swim faster, dive deeper and stay underwater for longer than any other bird. They are especially well built for aquatic life, with their dense feathering (in three separate layers), streamlined bodies and powerful flippers. They easily outperform all other birds in this medium.

The daddy of all penguins is the emperor, which lays claim to two astonishing records. Firstly it can hold its breath underwater for 18 minutes, far longer than any other bird. Secondly it goes down much further than the rest. Indeed, its average depth of dive when feeding is about 660 feet (200 m), which is deeper than the maximum depth recorded for any bird other than a penguin (namely, a Brunnich's guillemot, *Uria lomvia*, which went down to 443 feet/135 m).

When it comes to the greatest depth, though, the emperor penguin almost defies belief. It has been recorded at a depth of no less than 1,740 feet (530 m). (The deepest human free dive, unassisted by flippers or weights, has been recorded at 284 feet/86 m.) This raises all kinds of questions.

Yes, it must be able to hold its breath. And yes, it must be able to resist the pressure changes and avoid the "bends," although how it does this is unknown. Thirdly it must be resistant to extremely cold temperatures. And finally it must somehow find its food—fish and squid—in the dark depths, although how it does this is still a mystery. Bioluminescence has been suggested, but the truth is that once the emperor goes down far enough it enters a domain that is, to us humans, the edge of the unknown.

Deepest dive

NAME **emperor penguin** *Aptenodytes forsteri*
LOCATION Antarctica
ABILITY diving deeper underwater than any other bird

Fullest bill

NAME **Atlantic puffin** *Fratercula arctica*
LOCATION North Atlantic
ABILITY carrying large numbers of fish in its bill

An Atlantic puffin holds a delivery of fish for its youngster prior to entering its 40-inch (1 m) long burrow. It could be an anxious moment—in puffin colonies there are often gulls, parasitic jaegers or even other puffins hanging around waiting to steal food.

Birds are as variable in their manner of food delivery as they are in most other aspects of breeding behavior. Some birds, including several of the puffin's relatives, bring in only one fish at a time. Insect eaters may deliver hundreds of flies, plankton eaters thousands of plankton. It all depends on how the birds find and carry back the "groceries," so to speak.

By any measure, the Atlantic puffin certainly goes for bulk buying. There are about five fish here, which is numerically about average for a load, the range of five to 10 being normal. However, at times, presumably when an adult has struck lucky, many more may be carried, probably more than for any other bird. The record is an incredible 62 small fish in a single load!

One reason why puffins bring in a lot of fish is that they tend to seek out schools. They don't catch all the fish at once, but they have a cunning way of holding onto those that they do. The edges of the colorful bill point slightly backward, keeping some fish in place by friction, while others are partly impaled on the mandibles, their bodies held crosswise by the tongue. This is why the fish are not forever in danger of falling out in transit.

Atlantic puffins are generous to their chicks. They bring in 1½–2¼ ounces (43–62 g) of fish a day, compared with 1 ounce (28 g) for the common guillemot and ¾–1 ounce (20–28 g) for the razorbill. They do it in style too.

Strongest stomach

NAME **bearded vulture** *Gypaetus barbatus*
LOCATION mountains of Eurasia and Africa
ABILITY dealing with a diet of bones

A bearded vulture stands with a bone in its bill, proof of its credentials as a vulture and scavenger. However, it isn't about to treat the bone as a sparerib, nipping off the remaining flesh and sinew and discarding the hard parts, as other vultures do. The bearded vulture is a scavenger with a difference: it eats the bones.

Remarkably, some 70 percent of the bearded vulture's diet consists just of bone and marrow, which makes this species one of the most specialized feeders in the avian world. Its stomach is strong in every sense. The chamber contains far more acid than is usual in the stomachs of birds, helping to break the bone down quickly. And the sides of the esophagus and stomach must be made of iron, because bearded vultures frequently swallow bones up to 10 inches (25 cm) long, evidently without internal injury.

Although a long bone is manageable, bone fragments are better. Accordingly, the bearded vulture has the unusual behavior of smashing bones by dropping them onto rocks. It performs this with extraordinary dexterity and accuracy; it can hit a rock 33 feet (10 m) square from a height of 490 feet (150 m) and, if anything goes wrong with the drop, has been known to dive vertically almost 985 feet (300 m) after a falling bone to retrieve it, catching it in midair. The areas where these birds like to drop bones, patchworks of flat rocks that may be used by generations of bearded vultures over centuries, are known as ossuaries.

All this might seem to add up to a lot of effort to secure a substance of dubious nutritional potential, but in fact the opposite is the case. The nutritional value of the diet is actually 15 percent higher than that of the predominantly meat-based diet of other predatory birds.

Choosiest eater

NAME **rufous-tailed jacamar** *Galbula ruficauda*
LOCATION Central and South America
ABILITY catching specific butterfly species in flight

A rufous-tailed jacamar perches in the forest middle layer, on the lookout for flying insects. Foraging jacamars have a tendency to hold their heads up at an angle like this, so that they are firmly focused on what is moving just above them. They will pursue their prey in midair before returning to a perch to eat it. Remarkably, they are so focused on aerial pursuit that they will completely ignore food nearby, even if it is crawling at their feet.

The jacamar's favorite prey is large-bodied insects, especially butterflies. However, the jacamar has shown itself to be extraordinarily choosy in what it takes for its meals. It has reason to be choosy because there are, for example, dozens of species of butterfly in the South American forests that are unpalatable. Jacamars soon learn to avoid them, but, if they catch them by mistake, the noxious insects are released unharmed, except perhaps for a snip in the wings.

Many unpalatable insects exhibit what is known as aposematic coloration: by being big, showy and distinctive, they pass on the message that they are unpleasant to eat. In addition, unpalatable butterflies fly relatively slowly and directly, sending a further coded message that they should be left well alone. The jacamar is so expert a hunter that, in many situations, it can identify these butterflies in flight and avoid catching them. (Unfortunately this also means they miss out on these butterfly's palatable mimics.)

Because it understands all the messages, the jacamar hunts a special kind of butterfly—cryptically colored and flying on a fast and unpredictable course. These butterflies are a challenge, but they are wondrously nutritious and make it well worth the time spent avoiding the easily identifiable junk food.

Most patient feeder

NAME **shoebill** *Balaeniceps rex*
LOCATION central Africa
ABILITY standing motionless for half an hour or more

You're not going to use a bill like this for anything ordinary. The aptly named shoebill of central Africa is a true specialist, feeding almost entirely on lungfish—big, sluggish fish of well-clogged, sheltered waterways. They are not easy to catch, being large and awkward to deal with, and it takes refinements of fishing technique, as well as of the bill, for the shoebill to be successful.

One of those refinements is the shoebill's extraordinary patience. In some ways its fishing mirrors the technique of herons, waiting by the waterside and eventually striking when prey comes near. But the shoebill takes the waiting much further, sometimes staying completely still for more than half an hour; a heron, and any other stealth hunter for that matter, would have given up long before that. Observers watching shoebills feeding often miss the strike, having passed into a kind of torpor themselves.

The strike, when it comes, is a real all-or-nothing affair; it is often described as a "collapse." The shoebill lurches head first at the fish, and the rest of its anatomy follows. With a bill 7½ inches (19 cm) long it scoops up a huge mouthful, frequently containing some of the lungfish's habitat as well—water, plants and all—and it may take some time before the hunter regains its balance. A lungfish constitutes an ample meal, and after feeding the shoebill can go for several days without food. In the life of this bird, it seems, a lack of impetuous hurry is the rule.

Longest fast

NAME **emperor penguin** *Aptenodytes forsteri*
LOCATION Antarctica
ABILITY going for up to 115 days without food

This emperor penguin chick looks well fed and healthy as it basks in the gentle rays of the Antarctic spring sunshine. But its path to this point, and particularly that of its parent, has been long and hard. It could be argued that no parents of any bird, and perhaps of any animal anywhere, go to the same astonishing lengths that emperor penguins go to to achieve the success encapsulated here.

Emperor penguins breed, for safety's sake, on sea ice around the continent of Antarctica, up to 125 miles (200 km) from the open water and therefore out of reach of predators. It is a harsh environment, made much more so by the fact that, alone among birds, emperor penguins breed right through the Antarctic winter. The reason is simple: the four months of the austral breeding season are not enough for these large birds to complete their breeding in time to send chicks out to catch the ice-free seas of early autumn. So instead they lay eggs at the end of the season, spend the winter incubating them and allow their chicks to fledge during the warmest part of the year.

It's an extreme strategy. The male incubates the egg between his feet and folds of skin, standing around in temperatures that may plunge to -40°F (-40°C) in the middle of winter, enveloped by the dark and buffeted by the howling winds. Huddling together to preserve heat, the males endure this nightmare for up to 66 days, until the eggs hatch. Then the females arrive to relieve the males and to feed the chicks. The females may, in turn, have to wait another month or more before another changeover takes place, in similarly unpleasant conditions.

Amazingly, while the male incubates the egg he cannot feed at all, so instead lives off his copious body fat. If you take the incubation period and then add in the long walk to the nest site from the sea, and then the same walk back to open water, the total time the male may have to fast is an incredible 115 days. Needless to say, this far exceeds the time any other bird can go without food.

This is a close-up of the life of a scavenger. It isn't pretty—there is gore and death and decay. However magnificent Andean condors are in the air, their veneer of beauty disappears rapidly when observed gorging on a carcass.

Although we frequently see snapshots in the lives of scavengers on television—vultures crowding around a lion kill, for example—it is important to realize that most of a scavenger's time is spent in searching, not in feeding. Andean condors, for their part, may travel 125 miles (200 km) or more from the nest in the course of a day, scanning the ground for large carcasses, and still end up empty-handed. They are thought to be able to go a week or so without finding anything to eat. The kind of sustenance they need—big dead animals—is randomly and widely distributed.

Thus is it understandable that when they do finally come across a food bonanza—a dead horse, deer, guanaco or perhaps a sealion or whale on the coast—Andean condors seek to gain maximum advantage. They binge eat, as fast and as much as they can. They have been known to take 4½ pounds (2 kg) of meat in a single sitting, more than any other bird.

But such bingeing can create problems. The birds eat so much that they cannot take off. This is fine if the replete condors have time to hang around on the ground, and often they do just that, for many hours. However, if they are disturbed by a potential predator they have no choice but to vomit up excess food until they reach a weight at which they can take off. This must make them sick, in every sense.

Biggest binge eater

NAME	**Andean condor** *Vultur gryphus*
LOCATION	South America
ABILITY	bulk feeding

Greediest berry eater

NAME **Bohemian waxwing** *Bombycilla garrulus*
LOCATION taiga belt of the Northern Hemisphere
ABILITY eating a lot of berries very quickly

A great number of birds throughout the world eat fruit—it is one of the most readily available of all natural foods. From the tropics, where fruit is abundant all year round, to more temperate regions, where in the form of berries it is most widely eaten from summer through to early spring, there are birds for whom fruit is a major component of the diet.

However, despite the universality of eating fruit, there are comparatively few birds that could be considered specialized frugivores—that is, with fruit dominating their annual diet. Most of these are tropical forest species, such as cassowaries, cotingas and manakins, which exploit the biodiversity of their habitat to sustain their year-round habit.

It is perhaps ironic, therefore, that the species known to have guzzled the most fruit in a day is actually the Bohemian waxwing, a bird of the cold taiga belt of the Northern Hemisphere. Over much of its range it is by far the most frugivorous bird, and, in Europe at least, it is the only species that can subsist on berries for long periods without having to add in protein-rich extras, such as worms or insects. Puzzlingly it has no obvious unusual modifications of its gut to cope with such a monotonous diet.

The waxwing may be a specialist, but its consumption is still impressive. In the winter it eats between 600 and 1,000 berries a day. It is a comparatively small bird but has an unusually wide 4½-inch (11 cm) gape to deal with the size variation in its favorite diet. A berry diet washes down well with a drink, and one of the waxwing's other quirks is that, when thirsty, it sometimes catches snowflakes in flight.

All around the world a great many birds incorporate mimicry into their songs, adding sounds copied from their environment. They mainly do so to embellish their repertoires, much as we might include quotations in an essay: a variety of additional voices improves and enriches the message.

In terms of choosing the "best" mimic of the ornithological world, several parameters apply. Most mimics make perfectly faithful copies that are indistinguishable from the original, so purity is hard to measure. In terms of the sheer number of species imitated, the Lawrence's thrush (*Turdus lawrencii*) is probably the winner. It is known to have imitated at least 173 species of Amazonian bird. But for sheer panache and variety it is hard indeed to beat Australia's celebrated lyrebird. The song is a curious medley of sweet notes, rattles and trills, often loud and carried far. It is delivered while the bird is displaying its famous lyre-shaped tail, as in the picture.

Scientists have measured the lyrebird's song and found it to be composed of about 70 percent mimicry. An individual of average ability imitates the songs of around 20 species of bird. But it is beyond this parameter that the mimicry becomes extraordinary, for the lyrebird also imitates a whole range of sounds other than birdsong. Over the years they are known to have included the creaking of tree limbs, the sound of a kookaburra (*Dacelo novaeguineae*) snapping its bill, the wing beats of a large bird, the rustling of feathers, a koala calling, the bark of a dog, the squealing of young foxes, an ambulance siren, a saw, violin, cornet, and camera motor-drive, a child crying, a pig being slaughtered, and the husky cough of a heavy smoker.

Best mimic

NAME **superb lyrebird** *Menura novaehollandiae*
LOCATION southeastern Australia
ABILITY mimicking all kinds of sounds

Noisiest call

NAME **three-wattled bellbird**
Procnias tricarunculatus
LOCATION Central America
ABILITY making perhaps the loudest call of any bird

From its high perch in the Central American montane forest, a male three-wattled bellbird blasts out an advertising call over the treetops. The extraordinary "bock!" sound, which recalls the noise of a hammer striking a block, is deafening, even when heard from the ground beneath the bellbird's tree. It is known to carry for at least a mile (1.6 km), which, for the range of pitch, makes it louder than any equivalent bird sound in the world. The calls of cranes and the low-pitched "booming" of such birds as the Eurasian bittern (*Botaurus stellaris*) and the kakapo (*Strigops habroptilus*) carry further but are never as ear splitting as the pounding of the three-wattled bellbird.

The delivery of the call is a performance, as you can see from the picture. The bird opens its mouth extraordinarily wide and, prior to singing, contorts its whole body as it inhales air. Then comes the explosion of sound, and the wattles—one hanging from either side of the mouth, the other from above it—shake with the effort. The bird closes its bill, hops triumphantly into the air and settles back down on its perch, turning in the process to face the opposite way.

Perhaps surprisingly, young bellbirds take some time to perfect what sounds like a simple call. They don't proclaim at all until about 15 months old, and it then takes a full five months of additional practice to produce something relatively respectable. Furthermore, the "bock" sound is only one part of the bellbird's repertoire. Various, much softer, sounds accompany the loud headline, including squeaks and whistles. These take even longer to learn and, intriguingly, individuals in different parts of the bellbird's range each sing their own dialects.

There cannot be too many human records that have stood since 1952, such is our thirst for recognition and the stretching of boundaries. The desire of people to get their names at the pinnacle of their sport or pastime seems insatiable. Birds, of course, have no record of achievements, although they are just as motivated as any competitor to beat their rivals in one particular event—the race for recognition by the opposite sex.

It so happens, though, that a peculiar meld of avian and human endeavor came together in 1952 to create a record that has never since been beaten. An ornithologist by the name of Louise de Kirilene Lawrence was challenged by a friend to do a "Big Day." A Big Day is usually devoted to spotting as many species as possible, but Lawrence had a more interesting idea—she would count how often her local red-eyed vireo sang during the day.

The red-eyed vireo's song is a simple but variable phrase—"cheer-o-wit, cheer-ee, chit-a-wit ... "—that sounds like that of a rushed American robin (*Turdus migratorius*). An individual will have about 40 different songs, the same song never following in direct succession. It is not, however, the variety of its output that makes the red-eyed vireo notable, but the fact that this species can sing for hours on end.

Starting at dawn, Lawrence began to count the songs. By 5 a.m. the bird had sung nearly 1,700 songs, and an hour later more than 3,800. At times it paused in its repertoire, especially to preen, but would frequently feed on the hop and carry on singing. Altogether it sang for 10 out of the 14 hours of light available on that May day. The total at the end was a staggering 22,197 songs. That's a lot of singing, and a lot of counting!

Most tireless singer

NAME **red-eyed vireo** *Vireo olivaceus*
LOCATION North America
ABILITY singing all day long

Sharpest hearing

NAME **barn owl** *Tyto alba*
LOCATION worldwide
ABILITY astonishingly acute all-round hearing

Owls might be famous for their nocturnal habits, big eyes and round faces, but, contrary to the folklore, they don't see very much better in the dark than we do. Their truly astonishing talent is their acute hearing. The barn owl, for example, is perfectly able to catch living, fast-moving food in total darkness without using its eyes at all. Experiments in sealed buildings have shown how a barn owl can home in on a squeaking, rustling mouse using hearing alone, with a margin of error of just one degree to the horizontal or vertical. This bird could hunt with its eyes closed, were it not for the possibility of hitting obstacles.

There are two exceptional features that give the barn owl this amazing ability. The first is easily visible in the picture: its heart-shaped face. The ears are situated behind the outer rim of the facial disks, and the ridge between the eyes, together with the stiff feathers on the disk itself, reflect high-frequency sounds toward the ear openings, magnifying the signal.

The second unusual feature concerns the ear openings themselves. Not only are they large, but the left ear opening is slightly higher up than the right one. On us this lopsided arrangement might look distinctly odd, but it would add an extra dimension to our hearing—as it does with the barn owl. To home in on a sound requires a computation of the fractional time difference a signal takes to reach one ear with respect to the other. The positioning of our ears gives us this ability in the horizontal plane but not the vertical. The barn owl, however, has efficient three-dimensional hearing, hence its extraordinary performance when light fades.

Keenest eyesight

NAME **golden eagle** *Aquila chrysaetos*
LOCATION widespread throughout the Northern Hemisphere
ABILITY seeing detail at great distance

Anybody who has seen raptors hunt will be impressed by their apparently astonishing powers of vision. At one moment they can be soaring so high in the sky that they are almost invisible, and the next they are swooping down, with talons open, to grab a struggling animal hidden from our view. There are plenty of well-authenticated records of golden eagles being able to see small mammals from as much as over a mile (1.6 km) away, for example, and rabbits at almost twice that.

Although not all scientists are convinced that an eagle's eyes are superior to ours in any way, some estimate that these birds can detect the same level of detail as we can at between two and eight times the distance. This power to make such fine discriminations, known as visual acuity, would be a powerful tool for birds that feed on small, well-concealed animals.

Without a doubt, however, an eagle's eye is different from a human's. For one thing, it is actually larger, with the ample pupil allowing a generous amount of light to enter. The eye is also tubular, and well-developed muscles control the curvature of the lens quickly and easily, "focusing" the image with speed and clarity. The eagle's retina also differs from ours in that it has a higher number of detail-resolving "cone" cells, and two (to our one) special concentrations of receptors known as "foveae." The latter are not directed the same way, and it is thought that, together, they give the eagle an exceptionally good perception of movement.

All these modifications suggest that the term "eagle-eyed" remains a fair description of outstanding vision.

Quite a few birds could lay claim to building the most elaborate nesting structures on Earth, but as a family the weavers are the undisputed champions. Males of these small, sparrowlike birds elevate the chore of nest building into an art form, interweaving strands of grass and other plant fragments into sublimely neat and well-designed creations. It is a task of great skill, and female weavers judge prospective mates on the quality of their work. Good efforts facilitate sex.

Typically weaver nests are suspended from the tip of an overhanging branch. Males often begin by stripping off nearby leaves so that snakes and other predators cannot approach unnoticed. Their first construction task is to weave their materials into a vertical ring, suspended like a large curtain ring, which forms the middle of the structure. On one side they build the egg chamber and on the other the entrance. In the case of the African masked-weaver (shown here), the whole process takes five days, during which the bird will manufacture several thousand weaves and simple knots. Remarkably, all it uses for the construction is its bill.

The weaving process is innate, but experience also counts for a lot. Young birds often make a mess of their first efforts and fail to attract a mate. Even a more experienced builder may need to construct half a dozen nests before impressing a female, who may not even come to look unless the male performs simple displays beneath the nest, fanning his wings seductively and spreading his tail.

Craftiest builder

NAME **African masked-weaver** *Ploceus velatus*
LOCATION southern Africa
ABILITY weaving intricate nests from grass fragments

The rarefied habitat of the alpine chough is well demonstrated in this photograph, taken high up in the German Alps. This species lives only in mountain areas and regularly breeds at dizzyingly high altitudes; in the Himalayas, for example, it frequently breeds at 16,400 feet (5,000 m) or more.

That, in itself, does not make the alpine chough a record holder—a number of species breed at similar heights above sea level. Neither does the alpine chough hold the altitude record for any bird: that belongs to a flying Ruppell's vulture (*Gyps rueppellii*) that was hit by a plane at 37,075 feet (11,300 m).

What does set the alpine chough apart, however, is its propensity for finding and eating food at exceptional altitudes. With its superb mastery of flight this bird can reach anywhere in the mountains, even when the weather is wild. It is also an omnivore, ready and willing to try anything for edibility.

Thus it is that the alpine chough has been known to follow human hikers everywhere they go, picking up scraps of presumably high-energy foods. On one occasion a small party of birds followed some mountaineers to a height of 27,020 feet (8,235 m), the current record for high-altitude foraging.

Highest forager

NAME **alpine chough** *Pyrrhocorax graculus*
LOCATION mountains of Eurasia and North Africa
ABILITY ascending higher to feed than any other bird

This bizarre structure, draped over the stark, dead branches of a tree in Namibia, would not win any prizes for architectural splendor, but as a functioning dwelling it is invaluable to its residents. Indeed, its attractiveness as a property is demonstrated by the fact that it is not only the builders that live here, but a host of other tenants too.

The nest is built by a sparrowlike bird known as the sociable weaver, a small species with a black bib and scaly pattern on its back. A collective effort, put together over many thousands of hours, it is not so much a nest as an apartment building. Although all the birds share a communal roof, the underside is divided into compartments, each of which houses a pair of birds. In all, over 500 birds may live in a large nest at any one time. The advantage of the nest being so big is that it remains cool in the summer and retains its heat in the cool season. The desirable residences are sometimes snapped up by other birds, too, including the predatory pygmy falcon (*Polihierax semitorquatus*).

Although not the largest nest in the world (that accolade belongs to the nests of some eagles, which are built up over decades), the dimensions are still impressive. Some structures have measured up to 13 feet (4 m) tall and 24 feet (7.3 m) across, and they may weigh as much as a ton. Not surprisingly, they sometimes prove too heavy for the structures upon which they are built, which include not just trees but also telegraph poles. On the other hand, a well-built sociable weaver nest may well last for more than 100 years, housing generations of weavers. So if there is an architectural award for longevity, perhaps the prize should be awarded here.

Biggest communal nest

NAME **sociable weaver** *Philetairus socius*
LOCATION southern Africa
ABILITY building an enormous communal nest

Most prolific breeder

NAME **Eurasian collared dove** *Streptopelia decaocto*
LOCATION Europe east to southeast Asia
ABILITY attempting to raise the highest number of broods in a year

A collared dove feeds its youngsters, provisioning them with a unique type of "milk" synthesized in its crop and then regurgitated. Pigeons are among the very few birds in the world (along with flamingos and the emperor penguin) to feed their young on such a product. Supremely nutritious (containing 19 percent protein and 13 percent fat), the milk helps the youngster to grow quickly and move along the collared dove's impressive production line.

Everything about a collared dove's breeding seems to be dedicated to churning out young. For one thing the eggs are not incubated for long—only 13–18 days, which is short for the size of bird. Secondly the milk enables the chicks to grow so fast that, within a couple of weeks, they are able to fly. In this culture of haste the youngsters are turned out of the nest well before they attain anything close to the weight of the adults. A third shortcut is to overlap the breeding cycle, so that, when the father may still be feeding fledged young, the female will already be incubating the next batch of two eggs.

Many people who live in temperate parts of the world are surprised when they see nests of pigeons or doves in the middle of winter. However, this is yet another unusual feature of the breeding pattern of these birds. In contrast to the majority of birds, most pigeons and doves do not exhibit a "refractory period," a kind of post-breeding hiatus in which the relevant organs regress to prevent inappropriate procreation. So there is nothing to stop them producing young all year round. Currently the collared dove holds the record for the most broods attempted in a year—nine. This is an impressive testament to the resilience and productivity of pigeons and doves as a whole.

Strangest way to cool off

NAME **wood stork** *Mycteria americana*
LOCATION warmer regions of the Americas
ABILITY cooling off using feces

A wood stork forages in an American swamp, leaning down into the water with its bill open to catch whatever comes by. This species has the fastest known reaction time of any hunting bird; if something edible—be it a fish, frog, crayfish or anything else—brushes past its bill, the mandibles snap shut within 0.025 seconds. Not much gets away and, as a result, the wood stork usually enjoys a copious diet.

In common with most members of its family (Ciconiidae), the wood stork occurs in warm climates, including the very south of the United States and northern South America. As a result it rarely gets chilly, but overheating is a common problem. Its patient, standing-still method of feeding doesn't help either, because the wood stork often works under the glare of the sun.

Storks may cool down in various ways, including seeking shade where they can, panting and ruffling their feathers to let the heat out. But it is for another form of cooling that they are most famous—or infamous. Known politely and more scientifically as "urohydrosis," this is the practice of squirting waste products, both urine and feces, onto the legs. We may recoil at the thought of such a thing, but to a stork the resultant cooling due to evaporation is a blessed relief on a hot day; it could even be described as an indulgent luxury.

Only a few other birds practice urohydrosis. The shoebill (*Balaeniceps rex*) is one, while the American vultures (Ciconiidae) constitute the others. Indeed, the shared behavior of storks and American vultures is such that these birds are thought to be close relatives.

Best architect

NAME **rock wren** *Salpinctes obsoletus*
LOCATION western North America
ABILITY using rocks in the construction of its nest

A male rock wren sings from a slab of petrified wood on a boulder slope in the American Southwest. These birds are well named, living in territories that give them access to rocks of any kind, where they feed from cracks and under boulders.

The rock wren's attachment to hard substrates is not just one of foraging. It also stretches to the construction of its nest. On the whole, the smaller birds of the world tend to deal in fine materials for furnishings—grass, moss, hair, feathers—for the simple reason that they are light, flexible and have a neutral temperature, and it is highly unusual to find any species using such bulky material as stones. But the rock wren is an exception—in a big way.

The rock wren starts by selecting a crevice of some kind for its nest site. This site is often deep and sheltered, for example, tucked well under a large rock. One might think that, with such a hard floor, a little bit of softness underfoot would be welcome but, no, the wrens use pebbles to make their foundation, one of the very few birds in the world to do so. Once this is in place, they will, somewhat grudgingly, add some grass and other soft construction material.

This is by no means the end of their labors. For reasons that are not understood, most rock wrens now add a "feature," using up to 50 substantial stones to form a "pathway" up to the nest. Each stone may add up to one-third of the bird's own weight. It's a lot of effort and must serve some purpose, but for the moment only the rock wren knows what that is.

Smartest transporter

NAME **rosy-faced lovebird** *Agapornis roseicollis*
LOCATION southwest Africa
ABILITY transporting nest material in its feathers

Members of the parrot family are not, on the whole, famous for their nest-building prowess. This is actually an understatement—many of them don't make a nest at all, simply appropriating a hole in a tree or bank and performing a few housekeeping duties. Some species dig out their own hole, while others might chew the wood at the bottom of a cavity to make the interior a bit more comfortable.

The lovebirds of Africa, however, are an exception to this rule. Not only do they build a nest—and an intricate one at that—some species actually transport their nesting material in a way that is unique among birds.

This rosy-faced lovebird is one of the better-known species. It breeds in colonies, either in rocky crevices or the eaves of buildings, or in the communal nests of weaver birds. In the former instance the female constructs a neat cup inside the cavity using long strips of bark, leaves and coarse grass. She incises the materials from plants with her strong, sharp bill, then, when she has fashioned what she needs, she tucks the material into the feathers around her rump and lower back and flies to the nest hole relatively unencumbered.

Fascinatingly, other lovebirds vary in the way they transport their material. The related black-winged lovebird (*Agapornis taranta*) tucks its load anywhere in the plumage, whereas the yellow-collared lovebird (*A. personatus*) carries it in conventional style in its bill. Why this unusual, but convenient, feather-tucking behavior has only arisen in a small group of African parrots is a mystery.

Best thermal engineer

NAME **malleefowl** *Leipoa ocellata*
LOCATION Australia
ABILITY constructing an incubator for its eggs

A malleefowl flicks some sand behind it, using its large, powerful feet. At first sight this looks like a simple case of scratching the ground for food, kicking away excess litter to reveal tasty morsels beneath.

However, the mound upon which this malleefowl is standing is in fact its nest, and the kicking away of soil is part of a complex program of temperature regulation. The malleefowl and its kin are the only birds in the world that do not warm their eggs by applying body heat. Instead, with great labor, they build their own incubator and keep their eggs viable through a combination of solar radiation and heat generated by bacteria within rotting vegetation.

To pull off this feat of external incubation is not easy. First of all the nesting malleefowl have to build their mound, which is no small task. They begin by excavating a hollow a yard (1 m) or so deep and 10–13 feet (3–4 m) in diameter. Both birds then fill the trough with leaf litter, twigs and bark, laboriously kicking their material in from a radius of 65 feet (20 m) from the nest. They then dig an egg chamber in the top and, during a brief hiatus, allow the rain to moisten the mound and give a boost to the composting of the vegetation.

Once the female has begun to lay the first of her 20 or so eggs, the bulk of the work now passes to the male. For several months he will need to regulate the temperature of the mound by adding material or removing it, either of which may take hours every day to complete. Early on the warmth is driven by the rotting process, but in the summer the sun determines the mound keeper's schedule: at daybreak he scoops out sand to let the sun's rays heat the interior, only to fill it in later for insulating purposes when the day warms up.

In order to check the temperature the male malleefowl has a special trick. He inserts his bill into the pile up to his eye, using the interior of his mouth as a thermometer. This method is sensitive enough for the male to keep the temperature at around 91°F (33°C) for months on end, warm enough, but not too warm, for the eggs to thrive.

Longest life

NAME **royal albatross** *Diomedea epomophora*
LOCATION New Zealand waters, subantarctic
ABILITY longest-living bird recorded in the wild

Seeing this royal albatross tend its chick, the generation gap is not difficult to spot. In many small species of bird nestlings will be only a year or two younger than their parents, but in long-lived birds, such as albatrosses, there are clear generations—known as "cohorts"—within the population.

This young royal has only a 30 percent chance of surviving its first year of life, but once this tricky period has been negotiated, it can look forward to growing older than almost any other wild bird. Albatrosses as a whole regularly pass 40 years of age, and the oldest recorded wild bird was a royal albatross. Nicknamed Grandma, this individual lived on Taiaroa Head, on New Zealand's South Island. It was ringed as a breeding adult in November 1937 and was last seen, together with a chick, on June 12, 1989. Assuming that it took some 10 years to reach breeding age, it must have been about 60 years old at the time.

Based on their annual mortality rate of 3–11 percent, it is predicted that some albatrosses could live to 80 years old. However, this is without taking senility into account. In common with human beings, albatrosses grow old and more vulnerable, and their death rate sharply increases after the age of 30 or more. "Grandma," clearly, was a remarkable exception.

Most remarkable immunity

NAME **Japanese white-eye** *Zosterops japonicus*
LOCATION East Asia
ABILITY high immunity to disease

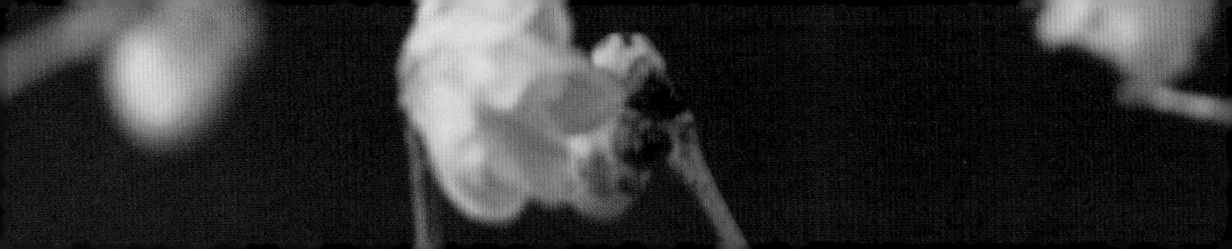

Extreme

Behavior

Oddest food fad · Biggest communal hunt · Cleverest hunter · Best fly fisher · Worst scrounger · Most voracious appetite · Funniest forager · Strangest "incubation" · Smartest scrounger · Dullest diet · Largest food store · Most bloodthirsty bird · Most persistent hitchhiker · Silliest antics · Craziest suitor · Longest copulation · Oddest time-share · Loveliest display · Shortest time on breeding grounds · Most impressive threat display · Best karaoke · Laziest feeder · Best avian soap opera · Biggest colony · Largest roost · Best tap routine · Best drummer · Most informative song · Longest journey on foot · Silliest migration · Shortest migration · Longest sleep · Most devoted ant follower

At first glance this parent great crested grebe appears to be mistakenly trying to feed its youngster with a feather—hardly the most nutritious of items one would have thought. But the grebe isn't erring. It is delivering feathers on purpose, and the youngster will accept the offering like a human child grabbing an ice cream. For grebes, feathers form an essential part of the menu. They eat thousands of them, mainly those molted from their own flanks and bellies.

Why on earth should they indulge in such an odd eating habit? The reason seems to have to do with the internal workings of the bird's gut. The feathers build up in the stomach and soon break down into a green paste, which may account for half the volume of the stomach. The paste plugs the entrance of the small intestine, preventing any unwanted hard substances, such as fish bones, from getting through. It might also be that the paste wraps up sharp objects and prevents them from damaging the walls of the gut.

Furthermore, the mashed feathers also enable the grebe to form pellets, which are coughed up through the bill. Whenever a grebe eats feathers it usually drinks large quantities of water too, and it is highly likely that, in the course of this coughing up, the grebe has the opportunity to rinse out its stomach contents, so to speak. This may allow it to get rid of any buildup of internal parasites, such as tapeworms and flukes, to which grebes are particularly prone.

Oddest food fad

NAME **great crested grebe** *Podiceps cristatus*
LOCATION much of Europe, Africa and Asia
BEHAVIOR eating large numbers of feathers

Biggest communal hunt

NAME **guanay cormorant** *Phalacrocorax bougainvillii*
LOCATION west coast of South America
BEHAVIOR cooperative fishing

A seething mass of guanay cormorants rides in the surf on the wild, rugged coast of Peru. One of the most abundant of all cormorants, the guanay is famous for being one of the "guano birds" of the Humboldt Current, present in such numbers that fertilizer from its droppings, along with that of the Peruvian pelican (*Pelecanus thagus*) and the Peruvian booby (*Sula variegata*), once constituted one of Peru's largest industries. Nowadays, thanks to overexploitation, the significance of the industry has drastically declined.

In common with other cormorants, the guanay pursues fish underwater for its food. It is heavily dependent on one species, the anchoveta (*Engraulis ringens*); if it is in short supply during the breeding season the birds will produce few young. The anchoveta is a deep-water schooling fish usually found far from land. From breeding sites on the rocky coasts the guanays commute to and from their fishing grounds, often in large numbers.

Although catching fish is a somewhat solitary business, guanay cormorants will sometimes use their numbers effectively, fishing communally. Several thousand birds may synchronize their dives, plunging in all at once and then surfacing in near unison. With so many birds on the hunt all at once, panic ensues among the fish.

At other times the cormorants will swim along the surface in a long line, controlling the direction of the fish and picking them off as they go. Needless to say, these mass expeditions are an impressive sight and constitute the largest cooperative hunting effort seen among birds.

Despite appearances, this black heron isn't being camera shy. Instead it is using a remarkable fishing technique known as "canopy feeding." It has effectively made a parasol out of its wings and is working in its own shade.

Although herons as a whole are well known for their patient standing and waiting, the black heron never maintains this posture for long. Instead, a foraging individual will walk across a shallow pool, put up its canopy for a few seconds and then move on to a new position, choreographing a distinctive stop-start ballet over the surface of the water. While holding its wings thus encircled it often paddles its feet within its own shade, presumably to flush prey. For a heron, this bird is something of a fidget.

What could be the reasons for striking such a posture? Well, as far as this hunter is concerned, working in its own shade reduces glare and reflection, and it makes the fish or shrimp at its feet easier to see and thus to grab. Also, black herons live in hot countries where, feeding in the open, they could easily be affected by the strong light. People find sunshades useful, so why not herons?

However, the main reason for canopy feeding is probably that it draws fish to the hunter. Fish naturally feel vulnerable in open, shallow water and habitually gravitate toward shelter and shade. Unfortunately for them, this sensible tendency can, beneath the black heron's canopy, prove to be a fatal attraction.

Cleverest hunter

NAME **black heron** *Egretta ardesiaca*

LOCATION sub-Saharan Africa

BEHAVIOR catching fish under its wings, so to speak

Adopting an almost horizontal posture, a green heron prepares to lunge toward a fish beneath a lily pad. Within seconds it will make a grab—aided by a special arrangement of the vertebrae that enables the neck to "snap" forward—and will soon have the fish wriggling in its jaws.

All over the world, other species of heron practice the same technique, and at first sight there seems nothing unusual about what this green heron is doing. There is, however, a very big difference. In many ways the green heron has more in common with a human angler than any other species. It regularly uses bait to lure the fish within its reach. Using tools is a rarity among birds, but the green heron, along with several very close relatives in other parts of the world, is known to place a variety of different objects into the water in front of it then lunge at fish attracted to the lures.

Green herons have been recorded using sticks, feathers, bread, captured insects, flowers and corn, all of which have been picked up away from the water's edge and dumped onto the surface. American birds have even been seen to use popcorn. Clearly some of these are edible items, but feathers and sticks act as "dry flies" of no nutritional value that are sufficient to fool a fish. Interestingly, some green herons snap twigs off waterside vegetation to produce their bait, a rare example not just of using tools but of toolmaking, too. Thus, in addition to the famed patience used by other herons, green herons employ cunning.

Best fly fisher

NAME **green heron**
Butorides virescens
LOCATION North America
BEHAVIOR catching fish using bait

Worst scrounger

NAME **parasitic jaeger**
Stercorarius parasiticus
LOCATION coasts and tundra of the Northern Hemisphere
BEHAVIOR living off the spoils of other birds

Getting this close to a parasitic jaeger is an intimidating experience, whether you are a human who has strayed too close to its nest or a seabird crossing its path. Jaegers are the only birds in the world to have claws on webbed feet, which, together with a sharp, predatory bill, means these birds don't just look dangerous, they are armed too.

Although jaegers can be killers, it is as scroungers that they pose the greatest menace to other birds. Scrounging to them is not a peripheral activity but a career path. Over the course of a year, a parasitic jaeger will obtain more of its food by theft than by any other, more "legitimate," means. Indeed, at times during the breeding season it uses food piracy to the exclusion of everything else; its diet, and that of its chicks, consists entirely of stolen goods.

Along with their built-in weapons of destruction, parasitic jaegers are superb fliers with long, pointed wings and slender bodies. This enables them to do two things. Firstly they can ambush and then hound a bird that is heading back to its colony with a beakful of fish, "tailgating" it until, exhausted and distressed, it drops its hard-earned meal. Secondly their expertise allows them to field the dropped fish quickly, either snatching it in midair or gathering it at the surface of the water before returning with the fish to their own territory.

So dominant is food piracy among parasitic jaegers that they time their breeding season to coincide with that of one of their main victims, the Atlantic puffin (*Fratercula arctica*). Other frequent victims are the black-legged kittiwake (*Rissa tridactyla*) and the Arctic tern (*Sterna paradisaea*).

Todies are minute birds found only in the West Indies. Related to the kingfishers, they share the same vivid colors and habit of building a nest in a tunnel. With their perky character they are popular and distinctive.

The Cuban tody pictured here is in typical pose. It looks gentle, almost cute. But this tody, like the other four members of its family, is actually a voracious predator. Its destruction of the small invertebrate fauna where it lives is extraordinary, and it has one of the fastest feeding rates of any bird in the world.

In several studies of todies conducted on various West Indian islands, researchers have been so struck by the birds' fabled appetite that they have collated statistics on their feeding. It appears that, within the 14-hour feeding time afforded by the tropical day, a tody catches between 1.1 and 1.9 items every minute, be they flies, moths, bees, ants or grasshoppers. That adds up to between 924 and 1,596 food items a day and constitutes 40 percent of the bird's body weight.

In the breeding season the toll on prey becomes even greater because todies acquire extra mouths to feed. Todies usually have three eggs, and during the late nestling stage they deliver food at a higher rate per chick than any other parent bird: an average of 140 feeds a day. One family in Puerto Rico is estimated to have consumed 1.8 million insects in five and a half months. Small insects beware!

Most voracious appetite

NAME **Cuban tody** *Todus multicolor*
LOCATION Cuba
BEHAVIOR fastest feeding rate among both adults and young

There are certain accounts of bird behavior that are so bizarre, so unlikely or so prone to cause mirth that it is difficult to accept upon first reading them that they are true. Take this story from the rain forests of eastern Australia concerning an otherwise unheralded species known as the bassian thrush, a retiring character of the forest floor. The story is well attested but, if you don't believe it, that's perfectly understandable.

Bassian thrushes feed almost exclusively among the leaf litter, where earthworms form their primary diet. There is nothing unusual about this—worms are a food item common among thrushes. The first sign that something strange happens came when observers noticed that, more often than usual in other birds, these thrushes regularly placed their vents close to the ground while feeding, as if squatting down to defecate. Furthermore, while this happened the researchers began frequently to pick up a quiet sound like an outward rush of air.

Yes, you've read it correctly. It seems as though bassian thrushes intentionally fart while they are feeding. The sudden jet of air is thought to disturb the earthworms, causing them to contract their bodies and give away their location; once this happens the birds can quickly turn around to grab them.

That is not the end of the story. On watching even more closely, the same observers found that, just before squatting, the birds would make another, gentler sound, equivalent to the gulping of air. With their short guts it appears that the birds can take in and break wind in quick succession, thus forming a mechanism for their astonishing method of foraging.

Funniest forager

NAME **bassian thrush** *Zoothera lunulata*
LOCATION eastern Australia
BEHAVIOR feeding by flatulence

Strangest "incubation"

NAME **boreal owl**
Aegolius funereus
LOCATION circumpolar boreal forest
BEHAVIOR reheating frozen food

In winter in the far north you have good hunting days and bad hunting days. This boreal owl seems to have had a good day. It hasn't yet eaten the red-backed vole in its talons, so it probably has a full stomach from earlier successes.

No matter. If hunting conditions are good the owl won't stop. It will carry on and on. A rapid change in the weather could soon put a stop to hunting, perhaps for days, but any surplus meat can be stored. The owl caches the small bodies of its prey in holes in trees or other crevices, sometimes lodging them in the fork of a branch. Five or six bodies may be collected together, a valuable resource for a rainy day.

In the winter caching is ideal because conditions keep the food fresh by freezing it. In theory the store could last for days, even months. There's only one snag, though, and that is that frozen food is not easy to handle or digest. Owls usually eat their food when it is still warm, so frozen bodies are hardly ideal for them.

So what do they do? Well, they have a ready-made solution. Birds warm their eggs by applying their body heat to them, a process known as incubation. And that is exactly what the boreal owl does. It squats upon the food to defrost it. A dose of heat soon makes a vole's body that much more manageable and tasty.

Smartest scrounger

NAME **greater honeyguide**
Indicator indicator
LOCATION sub-Saharan Africa
BEHAVIOR intentionally leading humans to bee colonies

A greater honeyguide perches on top of a branch, making itself as conspicuous as possible. This is a bird with little fear of humans, least of all photographers, and is not one to skulk away. Instead it sees people as potentially useful—not just the trappings of people either. This bird intentionally communicates with our species for its own ends.

The greater honeyguide's staple foodstuff is beeswax. Along with its kin, it is among the very few birds in the world to cope with this product, digesting it with special enzymes in its gut. In the African bush bee nests are plentiful but opening them up can be hazardous. Physically it is fairly arduous, but the real danger comes from the bees, which understandably do not take kindly to having their nests plundered. Although honeyguides have a specially thickened skin to protect them, they are vulnerable to bee venom. Individuals are sometimes found dead below the nests; one such victim was discovered with 300 stings on its body.

For a honeyguide, therefore, it is better and safer to get a bigger animal to do the dirty work, and that's where humans come in. People like honey and frequently raid bee colonies. Over the years, greater honeyguides have come to recognize this and have developed a symbiotic relationship with hunter tribesmen. The birds are better at finding the nests than people, and they have learned to guide humans to them—hence their name.

A leading honeyguide calls to a person with a loud, almost continuous chattering and flies around, flashing its conspicuous tail feathers. As the person follows the honeyguide continues in its excitable way until the bee nest has been reached, then it falls silent. After extracting the honey the humans traditionally leave the honeycomb for the honeyguide as a way of payment.

Dullest diet

NAME **western capercaillie** *Tetrao urogallus*
LOCATION Eurasia
BEHAVIOR surviving for months on a single type of food item

This western capercaillie stomps off through the snow, having just taken out its aggression on a photographer. These birds are notoriously irritable in the early breeding season, inclined to attack anything large that moves, be it a deer, a bicycle or, especially, another capercaillie. Their proverbial grumpiness is doubtless due to a spring dose of territorial pique. But capercaillies and other grouse will have endured many hardships during the preceding winter, so it is reasonable to allow them their moment of ire.

How could you, for example, cope with a winter diet consisting almost entirely of pine needles? That is what capercaillies have to put up with. Many individuals spend the entire season in a single tree, eating nothing but these leaves and a few shoots, and just sitting around for hours on end. Grouse have long intestines in which the bacterial flora break down the cellulose in the fibrous needles, extracting as much goodness as possible. But the lack of variety, though convenient, doesn't make the birds dynamos of avian energy.

Capercaillies are not the only ones who subsist this way. In North America, spruce grouse (*Falcipennis canadensis*) have a diet that may be 100 percent conifer needles, while greater sage grouse (*Centrocercus urophasianus*) depend to the same extent on sage, and ruffed grouse (*Bonasa umbellus*) graze monotonously on aspen buds. The red grouse (*Lagopus lagopus scotica*) of Britain and Ireland fares even worse: 90 percent of its annual, and not just winter, diet is composed of a single kind of plant, heather.

Largest food store

NAME **acorn woodpecker** *Melanerpes formicivorus*
LOCATION western North America south to Colombia
BEHAVIOR hoarding tens of thousands of acorns

It's one thing to make holes in trees, but this woodpecker seems to be overdoing things a bit. The species in this picture, however, is an acorn woodpecker, a bird renowned for its unusual hoarding behavior. In the northern part of its range it depends heavily on acorns; these may constitute 50–60 percent of its annual diet, and a much higher proportion in winter.

Acorns, of course, are a highly seasonal crop, so the woodpecker harvests them in fall and stores them away. The holes in this tree, as you can see, are used as small deposit boxes. The nuts, which may also include almonds, walnuts and pecans, are firmly wedged in to keep out thieves, those without strong bills, who might be tempted to help themselves when the woodpecker isn't looking.

The so-called "granaries," which may contain 50,000 holes drilled over the generations, are, however, a considerable draw to rivals. So, in order to protect a large and vital resource, acorn woodpeckers form groups that live together on a permanent basis, sharing the granary stores. During the breeding season, the groups, which may contain up to 12 adults, also nest collectively. Every member of the group helps to incubate the eggs and feed the young in the nest.

Most acorn woodpeckers use trees for their stores, but other wooden structures, including telegraph poles, may be used instead. This can be something of a headache for telephone companies, but for the woodpeckers the strong, smooth wood is ideal for their needs.

This distressing scene shows two Hood mockingbirds approaching an injured Nazca booby (*Sula granti*), as if lining up for the kill. In fact, the smaller birds are not as predatory as they seem, despite their long, curved bills and beady eyes. They are merely being opportunistic, attending the scene of an incident and hoping to profit from it, like ambulance-chasing lawyers.

It is not the mockingbirds that have caused the injuries. The young booby made the mistake of straying from its nest, and its horrific injuries were caused by attacks from neighboring adults. The mockingbirds, however, are certainly not there to help. Years of isolation on the island of Española (also known as Hood Island) have taught them to gather food from any source they can, blood included. They will drink the blood from this booby's wounds for as long as the injured bird is too weak to fight back.

Hood mockingbirds take blood and body fluids from other sources too. They gorge themselves on the placentas of sea lions and, most audaciously, peck the wounds of sea lion bulls wounded in battle. Human visitors are sometimes shocked to be followed by small parties of mockingbirds attracted to scratches on their legs. Drinking blood can be a risky strategy though. Animals don't take kindly to it, and there are many instances of mockingbirds being killed by the animals they are overzealously exploiting.

Most bloodthirsty bird

NAME **Hood mockingbird** *Nesomimus macdonaldi*
LOCATION Galapagos Islands
BEHAVIOR drinking blood from wounded animals

Most persistent hitchhiker

NAME **red-billed oxpecker** *Buphagus erythrorhynchus*
LOCATION eastern Africa
BEHAVIOR hitching a ride and snacking at the same time

This photograph shows a red-billed oxpecker in its natural habitat—on the head of a mammal! A member of the same family as the European starling, this strange bird is hardly ever seen away from the hides of large four-legged grazing mammals. It literally makes a living off other animals' backs.

Working on skin and fur, the red-billed oxpecker has unusual dietary preferences, verging on the vampirish. It is one of the very few birds in the world to specialize in eating ticks, which it gathers by using its bill like a pair of scissors to pry them off its host's hair or skin. If it wants a change it will also eat lice and various bloodsucking flies.

The common denominator in all these tasty morsels lies in the prey's diet—blood. The suspicion therefore arises that the oxpecker's taste is not for the protein in its prey's bodies, but the blood of its host. This is confirmed in a couple of ways. Firstly oxpeckers prefer engorged ticks to other ticks—they will eat 100 of the former a day—suggesting that the blood is the attraction. Secondly they often feed at animals' wounds, lapping at blood and body fluids without a tick in sight.

Certain animals act as preferred hosts. Zebras rarely top the poll, whereas giraffes, rhinoceroses and buffalo are very important, with warthogs a popular niche option. Not surprisingly the oxpeckers usually select large, tall animals, but certain hosts such as impalas, with a particular tendency to harbor parasites, buck the trend. Oxpeckers rarely forage on the back of elephants. Apparently these great beasts have particularly sensitive skin and cannot bear the touch of the oxpeckers' claws.

Silliest antics

NAME **kea** *Nestor notabilis*
LOCATION New Zealand
BEHAVIOR playing

The intelligence of parrots is well known, with the birds' ability to learn human sounds, for example, leading to their enormous popularity as pets and attractions at wildlife parks. Less celebrated is their capacity for play, which is well developed and perhaps reaches its pinnacle in the kea, a high-altitude species from New Zealand. In the tough environment where the keas live, adapting to the world needs to start early in life, and play is a good way of learning about relationships, foraging and survival.

Hikers in New Zealand won't necessarily be impressed. They are sometimes persecuted by gangs of keas stealing their food and fiddling around with anything left unattended. Keas have been known to steal cups and mugs and throw them downhill like mischievous adolescents. To make matters worse, they have been recorded sliding down the sides of tents early in the morning, waking the inhabitants up.

Sliding is, in fact, a common pastime. In the cold mountains where they live, keas regularly tumble down snowy slopes, sometimes on their back. They also slide down windshields and roofs and, in captivity, have been known to make snowballs. Such play is usually conducted in groups of all ages, and the members frequently play fight and just mess around. Other antics include lying with their feet in the air and kicking at individuals on top of them and even swimming on their backs.

In a gesture of togetherness, a male superb fairywren treats its mate to a spot of mutual preening. The closeness of the two birds is not just physical; fairywrens keep their mates for life and reside with them in the same territory all year round. The familial bliss is added to by the youngsters, which often remain with their parents for a year or more. Encountering fairywrens in the wild, it is normal for an observer to see a small group of them.

Nevertheless the relationship between this pair could be described as "open" because, although the two birds live together, their sexual relationships frequently stray outside the pair-bond. Both sexes can be promiscuous; indeed, the superb fairywren is one of the least "faithful" of all birds to its primary relationship. In a study it was found that only 24 percent of eggs laid by a female fairywren were fathered by her social mate. The reasons for such a high rate of extra-pair copulation are unclear.

The behavior of the female is particularly intriguing. During her fertile period in the breeding season she makes regular predawn visits to copulate with various males. The male is equally accommodating to females that might visit him during the same time of day.

The seed of this twilight activity is sown in the months before breeding begins, when male fairywrens make regular visits to neighboring territories, displaying their wares. They are often decidedly indiscreet, displaying to the incumbent females even when the territory-holding male is present. Hilariously, they will even bring gifts of flowers in their bills, and these are always yellow. It is on such meetings, it would appear, that the extra-pair copulations are planned. So the female fairywren's gesture of conjugal fidelity seen here conceals a rather more promiscuous reality.

Craziest suitor

NAME **superb fairywren** *Malurus cyaneus*
LOCATION southeastern Australia
BEHAVIOR male bringing flowers to his "mistress"

A fully paid-up member of the fraternity of small brown birds, the aquatic warbler hides a surprising lifestyle beneath a deceptively demure exterior. Its pairing system is unique, involving an unusually prolonged union between male and female that could be the longest of its kind in the world.

Aquatic warblers breed only in Europe, in a scarce and declining habitat—marshes with clumps of sedge up to 32 inches (80 cm) tall with water in between. Such a habitat is exceptionally rich in insect life, a factor that is thought to have emancipated male aquatic warblers from any kind of breeding duties. The female builds the nest, incubates the eggs and rears the young on her own, while the male simply sings and seeks copulation.

Not surprisingly, with so little else to do, the males expend a great deal of effort on ensuring their paternity. They have multiple partners, and their testes are larger than usual for the size of bird, suggesting that sperm competition is fierce. They also have another extraordinary strategy to ensure that it is their genes that prevail. In the spring evenings, when the females are most fertile, males seek them out and, presumably to keep other males away, simply remain on their back for an astonishing length of time. While in most birds copulation lasts less than a second, in the aquatic warbler it can go on for half an hour.

The strategy is not as definitive as it sounds, however. Most clutches of six or so eggs are sired by two males, and some by as many as four. In such circumstances the long coupling must be a necessity for ensuring any breeding success at all.

Longest copulation

NAME **aquatic warbler** *Acrocephalus paludicola*
LOCATION Europe
BEHAVIOR perhaps the longest-lasting copulation among birds

Oddest time-share

NAME **band-rumped storm petrel** *Oceanodroma castro*
LOCATION Galapagos Islands and Azores archipelago
BEHAVIOR two populations sharing the same nesting burrows

In the bird world, nest sites tend to be sacrosanct. Where possible, a good many species return to the same site each year to breed, sometimes coming back from thousands of miles away, homing in with complete accuracy. Once established, owners of sites don't usually take kindly to challengers and, where there is conflict, it can be bloody and final. A nest site stolen is a nest site violated.

This band-rumped storm petrel is doubtless like every other bird in this respect, claiming its site vigorously every year. But what is different in this case is that—knowingly or otherwise—it shares its burrow with another breeding pair of the same species. The two sets of owners, however, do not actually meet at the burrow. This is because they have different breeding seasons; in effect, they time-share. So far, two populations of band-rumped storm petrels have been found to do this. On the Galapagos Islands one shift lays its eggs in May or June, while the other breeds from November to January. Meanwhile, on Baixo and Praia in the Azores archipelago one shift lays in the summer (June to July), the other in fall (October to December). It also appears that, in the Azores at least, individuals do not switch from one shift to another.

Indeed, in the Azores population there is a very slight difference in morphology between the hot-season and the cool-season breeders. The "cool" birds are slightly larger and heavier than the "warm" ones, and their eggs are longer and heavier. The differences are minute, and the birds are indistinguishable in the field—to people and presumably to each other. Genetic studies, however, hint that these birds may be a few steps on the way to becoming separate species.

Loveliest display

NAME **Indian peafowl** *Pavo cristatus*
LOCATION Indian subcontinent
BEHAVIOR the well-known spreading of the "tail"

It is possibly so familiar that its impact is diminished, but is there any courtship display finer or more spectacular than that of the peacock? The shimmering fan of feathers with its "eyes" of iridescent green, blue and orange has been admired by humankind for centuries (we probably enjoy it more than the famously indifferent peahens). It provides proof that, among animals, it is worth risking conspicuous ornamentation to ensure successful sexual selection.

Male peafowl compete with each other for possession of territories that can be used as arenas for their finery. Territories are often close together, and collectively they provide a forum for females to come and check out the male "talent." The males make their famous crowing to ensure that they advertise their presence and spend much time waiting for female visitors to show up.

As soon as a peahen enters his arena the peacock turns to face away from her, presenting the dull-colored back of his fanned train. Then he shivers his wings and, little by little, tries to steer the female to the center of the arena. Once she is there the performer turns as if making an actor's entrance, noisily shivers his train and presents his colorful fan full on. To us it is an overwhelming blur of color and drama.

To peahens, however, the charms of the fan—which, incidentally, is not the tail but the feathers below it—are related more to quantity than impact. Male peafowl may have up to 150 "eyes" on their train, and it seems that it's a case of the more the merrier. A female visitor almost invariably copulates only with the male that has the most eyes on his feathers, which is usually the oldest and most experienced individual.

Peahens are very assiduous in choosing their mates. It is only after repeat visits to many males, often over many weeks, that a peahen will finally allow the lucky winner to mount her. It's hard not to imagine her coming along with a little notebook, carefully counting eyes—hardly the way to appreciate what must be considered the best courtship display in the world.

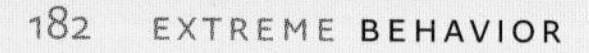

A curlew sandpiper tends its nest on the treeless tundra of the Russian Arctic. You don't have to look too closely to know that it is a female because male curlew sandpipers don't take part in looking after the eggs or young. Indeed, the father or fathers of this brood will have left the breeding grounds already.

The mating season of the curlew sandpiper is incredibly wild and short. Arriving by mid-June, the sexes don't form any kind of lasting bond. Males spend their tundra days pursuing every female in sight, in aerial chases, and it would seem that both sexes are promiscuous. The breeding system is thus cut down to its most basic and functional, and once copulation is over the males have played their part. Some individuals remain on the breeding grounds for as little as two weeks before heading south.

What is so extraordinary about this is the sheer effort involved in getting to the breeding grounds in the first place for this briefest of flings. While the curlew sandpiper has a restricted breeding range, concentrated on the Taimyr Peninsula and lying entirely within the limits of 70–128°E, it has a mighty wintering empire, encompassing coasts from west Africa to Australia. Some birds migrate more than 6,200 miles (10,000 km) between their breeding and nonbreeding areas. Indeed, the journey to their breeding site takes longer than their actual stay.

Shortest time on breeding grounds

NAME **curlew sandpiper** *Calidris ferruginea*
LOCATION Russian tundra
BEHAVIOR making brief visits to the breeding areas

At a superficial glance it's easy to see that this photograph depicts a colorful bird standing upon its nest, its wings spread wide. But imagine for a moment that you are a small creature moving along a branch in the dimly lit edge of a forest when, suddenly, a creature springs up in front of you which, judging from its huge, widely spaced eyes, must be enormous. Startled, you would doubtless follow your instincts of self-protection and scurry away in the opposite direction.

That is, of course, exactly what this sunbittern intends to happen with this impressive and spectacular defensive display. One moment it is a slow-moving, retiring, cryptically colored inhabitant of watersides that is difficult to see; the next it is a pretend monster that is hard to miss! This so-called "frontal display" can be held for a minute or more, and the bird sometimes swivels from side to side for extra effect.

But does it work? Well, to judge by observations and by the importance of this posture to the birds themselves, emphatically yes. Sunbitterns have been watched successfully fending off much larger birds than themselves, such as ibises and kites. Perhaps the most impressive and intriguing clue to its efficacy lies in the sunbittern's complete (and almost unique) lack of juvenile plumage. The bird molts straight from downy chick to costumed performer, and hence performs its display from early on, getting in a lot of practice.

Most impressive threat display

NAME **sunbittern** *Eurypyga helias*
LOCATION South America
BEHAVIOR spreading its wings in threat display

Best karaoke

NAME **purple sandpiper** *Calidris maritima*
LOCATION tundra of Eurasia and North America
BEHAVIOR pretending to be a rodent

A purple sandpiper sits tight on its nest in the tundra, its plumage blending in with the surrounding vegetation. This individual is relying on its camouflage to keep it and its eggs safe, hoping that predators will pass by without noticing anything.

If the worst does happen, however, and an Arctic fox or other enemy approaches too close, the purple sandpiper has a plan B: distract the predator's attention away from its vulnerable brood. The various displays that shorebirds use to do this are well known. Some, for example, pretend to have a broken wing, luring the predator away by the prospect of an easier meal. Other waders almost deafen the predator with noisy displays, while others actually make mock or real attacks.

The purple sandpiper's method is altogether more cunning. Sandpipers and other shorebirds share the tundra with large numbers of small mammals, especially voles and lemmings, which form the staple prey of most Arctic predators. Aware of this, the purple sandpiper simply pretends to be one. If it can convince an Arctic fox that there is a rodent in the vicinity, the purple sandpiper can lead the fox away from its nest.

Thus, when the fox appears out of the vegetation, almost upon its nest, the purple sandpiper fluffs out its feathers like fur, zigzags away like a startled rodent and squeaks, karaoke style, like a small mammal. The black stripe that splits its white rump is thought to imitate the black stripe on the back of a lemming, further adding to the deception.

If the ruse works the fox will certainly have a surprise when the "lemming" finally flies off, well away from the eggs and nest of that purple sandpiper the fox thought was somewhere in the vicinity.

Laziest feeder

NAME	**griffon vulture** *Gyps fulvus*
LOCATION	Eurasia
BEHAVIOR	getting up late and being lazy about finding food

This griffon vulture launches itself off its roosting cliff, setting out on an expedition to find a carcass. It is one of the more dramatic moments of its day; the rest of its time is mainly spent soaring up in the sky or just sitting on a cliff ledge in luxurious leisure.

To our perception, griffon vultures have an enviably short working day. They don't leave their roost until at least two hours after dawn, and they return between two and three hours before sunset. In inclement weather, or when they have gorged themselves the day before, they don't actually take off at all. What makes this even better is that they have an excuse. Griffons are large, heavy birds and, if they have to flap their wings for any significant length of time, they tire easily. They are thus dependent on harnessing the power of updrafts or thermals to keep themselves aloft.

Most Eurasian griffons live in mountainous areas, where winds are continually deflected upward and provide passage for the great birds. Griffons that live on the African plains are more liable to use thermals. These form when the sun heats the ground and, by proxy, the air above it; this warm air, surrounded by cooler air, then rises as a large bubble in which the rotating gusts can be very turbulent. The power of the rising air is more than enough to lift a broad-winged griffon. When the bird reaches its required height it drifts away from the thermal column and slowly glides down until, needing lift again, it catches another one. Thus the short days of griffons are filled with effortless rising and falling on the breeze. Hardly a bad life!

Best avian soap opera

NAME **white-fronted bee-eater**
Merops bullockoides
LOCATION south-central Africa
BEHAVIOR group living

Who needs to watch a television soap opera when all you have to do is visit a white-fronted bee-eater colony? The goings-on among this species have all the ingredients to enthral even the most committed couch potato.

Every soap is rooted in family squabbles, and the relationships among the bee-eaters are said to be more complicated than those of any other bird. The species is found in discrete colonies numbering up to 400 individuals, nesting in densely packed holes in sandy cliffs. The colonies, which constitute the bee-eaters' largest social grouping, are largely self-contained, with little or no social overlap with other colonies. Each colony is divided into clans—groups of 2–3 extended families whose members all share a feeding territory remote from the colony. The next social unit down is the extended family itself, which includes a pair, their young and auxiliary helpers who are generally male offspring from previous years. Down from this grouping is the pair itself, the official reproductive unit.

Just because the social structure is clearly delineated doesn't make it peaceful or free from strife. Far from it. In such a close-knit format sexual infidelity is rife. During breeding every male has to guard his female carefully in case of rape, which happens to one-fifth of all females; at the same time every responsible paired male is liable to succumb to the same temptation himself. Young males, who are supposed to be helpers, sometimes succumb as well. And, as for the females, their weakness is to dump eggs into the nests of nearby neighbors in order to have their peers unknowingly adopt their genetic material. In fact, 16 percent of all nests are parasitized in this way.

Criminal activity of one type or another enlivens most soap operas, and among white-fronted bee-eaters every colony has its bad "eggs" too. Certain individuals are career thieves, repeatedly stealing insects from birds arriving at the nest. One way or another, there is never a dull moment on the set of this particular soap.

The red-billed quelea holds the title as the world's most numerous wild bird. Despite being confined to sub-Saharan Africa, it beats hands down all the more familiar competition, such as the house sparrow (*Passer domesticus*). Its total population, estimated at a billion birds, tends to be heavily concentrated, and millions of birds breed together at any one time. So, as well as holding the world population record, its colonies are larger than those of any other bird.

Establishing exactly what the figures amount to, however, is not so easy. Quelea colonies are not like those of seabirds; they are not as spaced out, and the birds are smaller and therefore much more difficult to count. The queleas also form temporary aggregations that can spring up and disappear again within weeks and hence can easily be missed.

Nevertheless, good estimates of quelea numbers are still impressive. It is thought that a site covering 500 acres (200 ha), which is fairly typical, will hold some 10 million nests. There can be little doubt that some are larger still. A substantial tree within the site may hold 6,000 nests, all of them springing up in a week or so of frenzied building. Imagine, then, just how many nests would be packed into a well-wooded savanna of substantial size.

Not surprisingly, with so many birds so densely packed, queleas require plenty of food from local sources. Unfortunately they are seed eaters, and at times they will raid nearby crops, covering the plants like locusts and destroying everything. Thus, to add to their accolades of extremity, red-billed queleas also constitute the most serious avian pest, agriculturally speaking, in the world.

Biggest colony

NAME **red-billed quelea** *Quelea quelea*
LOCATION sub-Saharan Africa
BEHAVIOR forming enormous colonies

A swarm of red-winged blackbirds mills over the marshes at Bosque del Apache Wildlife Refuge in New Mexico. Some individuals have traveled as far as 50 miles (80 km) to be part of this vast roosting concentration, in which millions of red-wings, along with European starlings and other blackbirds, bed down every winter night, forming a black cloud in the whispering cattail stems.

The red-winged blackbird, with the male's jet black body and scarlet and yellow epaulettes, is one of the most familiar and abundant birds in North America, the population in the 1970s being estimated at 190 million birds. It is also sociable in the extreme. It breeds in colonies and forages in flocks, but its most impressive gatherings are at winter roosts. A roost in the Great Dismal Swamp in Virginia, for instance, is thought to contain 8 million red-wings and 7 million birds of other species, such as starlings and common grackles—probably the largest nonbreeding gathering in the world.

Sleeping in crowds, with all the noise, mess and hassle, might not seem to be the ideal way to spend the night and, indeed, the function of vast roosting gatherings in general is something of a mystery. Birds don't huddle together in body contact, and microclimate analysis shows that they don't benefit from additional warmth. The birds are not especially safe either: much as concentrations reduce each bird's chance of being predated, they also attract predatory birds to them. The most plausible explanation seems to be information gathering. Birds in close proximity in a roost can monitor their colleagues' state of health. If a needy bird roosts near a well-fed, healthy one, the former could follow the latter to its productive feeding grounds the following day.

Largest roost

NAME **red-winged blackbird** *Agelaius phoeniceus*
LOCATION North America
BEHAVIOR forming gigantic roosts

Best tap routine

NAME **red-bellied woodpecker**
Melanerpes carolinus
LOCATION eastern North America
BEHAVIOR tapping for multiple purposes

The bill of a woodpecker is a multipurpose instrument. It can be used for violent destruction one moment and subtle shaping the next and for territorial pronouncement in the morning and courtship in the afternoon. It provides its owner with the staples of life: food to eat, a place to live, somewhere to sleep and a means of communication. Together with the bird's keen mind, it makes woodpeckers the great success that they are.

This particular species, the red-bellied woodpecker of North America, is typical of the family. It can use its bill to obtain insects from just below the bark surface, employing an action, like a pickax, to obtain invertebrates that bore into wood and are out of reach of other birds. What many people don't realize, however, is that woodpeckers like the red-bellied often use their bills in more subtle ways first, tapping to detect where the bark is worn and where there may be hollow chambers inside.

Throughout the year the hard, heavy tapping is used to make cavities for nesting and roosting. It can take up to three weeks to create a single hole, but the protection afforded makes these residences uniquely desirable, not just to woodpeckers but to legions of other hole-nesting birds. The tapping is also put to good use to advertise for and attract a mate. The so-called "drumming" is not, as many people think, the sound of excavation of a hole, but it is actually the knocking of the bill against a suitably sonorous piece of wood simply to make a signature noise.

Best drummer

NAME **palm cockatoo**
Proboscige r aterrimus
LOCATION northern Australia and New Guinea
BEHAVIOR using a stick as a musical instrument

The magnificent wispy crest of the palm cockatoo gives the bird a distinctive look, which, with a little imagination, could easily be likened to the wild hairdo of a drummer in a rock band. For the palm cockatoo is the percussionist of the bird world, the only species to fashion a musical instrument and make a sound with it. The point of the performance is to impress a female, so the motivations of bird and man doubtless overlap here too.

If you want to hear the palm cockatoo performing you need to be in tropical Australia or New Guinea early in the morning or late afternoon, especially in the months of June and July. This is when the bird makes its way to the top of a dead eucalyptus, 30–50 feet (10–15 m) above ground, where it can find a nearly vertical trunk that has become hollowed out with age. This is the sounding board against which the cockatoo will play the beat.

Although it will sometimes use its feet to tap the hollow trunk, or even bang it with a nut taken from a nearby tree, the palm cockatoo customarily uses a stick for the best effect. Indeed, while the sound of foot tapping may be quiet, the click of a stick can be heard more than 330 feet (100 m) away. Thus, the cockatoo nips off a fresh branch about 1 inch (2.5 cm) thick, removes the leaves and bark and trims it to a length of no more than 12 inches (30 cm). It then brings this stick to its sounding board and, using one foot, bashes away, striking from 2 to more than 100 times.

This performance is unique in the bird world. Woodpeckers and other birds use their bills to tap, or their feathers to make vibrating noises in the wind. But no other species makes or employs such a complex tool to get its message across.

Most informative song

NAME **Eurasian blackcap** *Sylvia atricapilla*
LOCATION Eurasia
BEHAVIOR including coded messages in its song delivery

A male Eurasian blackcap sings with gusto from a bush in its territory. The pleasant whistled phrases are a well-known sound to bird watchers in Europe, and they carry a familiar message to other blackcaps—"keep out" to other males and "I'm here" to females—which is, essentially, the message of most bird songs. However, that's only a part of the narrative that this Eurasian blackcap is delivering. Recent detailed studies have shown that song output in males of this species gives away surprisingly detailed and useful information to listeners.

What the study found was that a Eurasian blackcap's song rate provides a faithful description of its territory. This is because males with a high song rate tend to have denser vegetation within their borders than less intense singers. Dense herbage hides the nest more effectively from predators than an open structure, and therefore the young are more likely to survive from nests situated in this kind of territory. So if a female monitors a male's song rate and finds it to be in the range of 160–180 songs per hour, she can expect to secure a safe breeding site and a better future for her offspring.

Many male Eurasian blackcaps, however, only sing in the range of 80–100 songs an hour. Their territories are less effective for concealing nests, and they tend to be less attractive to females. Nevertheless, there is a trade-off. These less able singers compensate for their shortfall by being better fathers—they give more assistance in incubation and make a higher contribution to feeding the young.

Longest journey on foot

NAME **emu** *Dromaius novaehollandiae*
LOCATION Australia
BEHAVIOR traveling large distances on foot

The emu is the second largest bird in the world, standing 5–6¼ feet (1.5–1.9 m) tall, about the same as a human being. Like the ostrich, the world's tallest bird, the emu is flightless and goes everywhere on foot. Normally it walks with sedate strides, but when pressed it can build up to a good speed, 30 mph (48 km/h) being the maximum recorded.

In Australia, where the emu lives, rainfall is often patchy and unpredictable, and it suits the emu, as it does so many other local birds, to live a nomadic lifestyle—to quote the local buzzword, it goes "walkabout." This means that populations do not usually make specific movements. Instead they wander in search of food supplies, staying put for some months if conditions are good, but always eventually moving on. On occasion these wanderings can be considerable. Birds have been recorded traveling 9 miles (14 km) per day and covering 335 miles (540 km) in nine months. This is almost certainly the longest migration by any bird on foot, since most other large flightless birds are more sedentary.

Studies have shown that emus converge on areas where rain has fallen, and it seems that the birds study patterns of cloud formations to help them in this regard. Cyclones attract birds from further away than do simple thunderstorms, because the cloud banks associated with them are larger and higher.

Not all emu migrations are random, however. In Western Australia there is a general movement southwest in winter and northeast in summer, following the rains, and no doubt similar patterns can be observed elsewhere.

Silliest migration

NAME **northern wheatear** *Oenanthe oenanthe*
LOCATION Eurasia, North America and Africa
BEHAVIOR using inexpedient wintering grounds

A young male northern wheatear seems to pause and ponder its journey ahead. Many northern wheatears breed in Alaska, and they have to travel a very long way— about 9,300 miles (15,000 km)—when returning to their winter home in Africa, halfway across the world. This journey passes through lands that would probably suit wintering wheatears perfectly well, but they don't visit them.

Neither do they make things easier for themselves by migrating southeast, where a southern flight of a mere 3,100 miles (5,000 km) would doubtless be sufficient to bring them to a suitable climate and habitat for the nonbreeding season in Mexico or elsewhere. But the wheatears carry on regardless and stubbornly refuse to change direction.

At the same time, across the continent, other northern wheatears are arriving on the tundra of eastern Canada and Greenland, having also wintered in Africa. Theirs is not such a long journey as that taken by Alaskan birds, but they still have to cross nearly 1,900 miles (3,000 km) of ocean between Africa and Greenland. Due south of their breeding grounds lie some pleasant lands in continental South America that look ideal for wintering. They are far closer, but the wheatears ignore them too.

Of course, the northern wheatear doesn't willfully ignore the more sensible option of wintering in the New World. It just isn't programmed to take it. Instead, its behavior is thought to be a superb example of how migratory journeys were affected by the last glaciation. During the last ice age, northern wheatears were almost certainly restricted to breeding seasonally on lands within easy reach of Africa. As the ice retreated the birds gradually colonized the lands opening up, spreading west and east as well as north. They would have spread little by little, stretching their longitudinal limits year by year, yet always returning to Africa as a refuge for the winter. Eventually, of course, they would have reached a point at which they were migrating unnecessarily long distances, but this would never have occurred to them.

Nowadays, with its migration route wildly over the limit, the northern wheatear remains stoically unaware of the easier paths it could be following.

Shortest migration

NAME **Clark's nutcracker** *Nucifraga columbiana*
LOCATION mountains of western North America
BEHAVIOR making minute seasonal movements

When people think of migration, they tend to imagine long journeys that take birds north, south, east or west. But there is a host of species for whom latitudinal movement is irrelevant—it's altitude that matters.

An example is the Clark's nutcracker of western North America, which usually breeds at between 5,900 and 8,200 feet (1,800 and 2,500 m) in altitude, often in woods at the outer edge of the tree line. This bird is a particularly hardy character that can breed very early in the spring, when heavy snow is still on the ground. However, it will not necessarily see out the winter at this altitude; in fall it sometimes retreats a few hundred yards lower down. By taking the edge off the winter cold, this relatively small shift may make all the difference to the bird's chances of survival.

Can this be considered a migration? Absolutely. Just because it is measured in yards rather than miles does not diminish the migration's effectiveness. All over the world, birds accrue the same advantage from minute downhill displacements that they would gain from several hundred miles of latitudinal movement. The climate is often radically different at the bottom of a mountain range in comparison to the top.

One of the curiosities of altitudinal migration is that in fall, when most birds in the Northern Hemisphere are moving south, some altitudinal migrants move north. The water pipit (*Anthus spinoletta*) is a good example of this, breeding in the mountains of Central Europe and wintering in Britain and the Low Countries.

A common poorwill—named after its soft, sweet "poor-will" advertising call—sits motionless among the rocks in the Arizona desert, its superb camouflage making it almost impossible to spot. However, the bird's coloration needs to be exceptionally cryptic. This member of the nightjar family is famous for its ability to "hibernate," remaining motionless and therefore highly vulnerable for long periods of time. It is the only bird in the world that sees out the winter period by sleeping.

Strictly speaking, the poorwill doesn't hibernate, which would involve chemical changes in the body. It simply goes torpid, allowing its metabolic rate to drop below normal so that it uses less energy. The poorwill, however, pulls off this trick to a greater extent than any other bird and for much longer too. While many species of bird, including other nightjars, are known to become torpid overnight, or perhaps for a day or two, the poorwill can settle into this state for as long as 100 days. In so doing it is able to avoid the winter cold, which diminishes its food supply—nocturnal insects—to almost nothing.

The torpid state reduces the poorwill's various bodily functions, including heart and breathing rates, and at the same time the body temperature drops. This drop can be spectacular, from a normal 104°F (40°C) down to an extraordinary 41°F (5°C), the latter by far the lowest recorded for any species of bird. Although verified by science only recently, the sleeping habits of the poorwill have been known for generations. Indeed, the local Native American name for this bird may be translated as "the sleeping one."

Longest sleep

NAME **common poorwill** *Phalaenoptilus nuttallii*
LOCATION deserts of North America
BEHAVIOR becoming torpid for weeks on end

Most devoted ant follower

NAME **ocellated antbird**
Phaenostictus mcleannani
LOCATION Honduras south to Ecuador
BEHAVIOR following ant columns

The ocellated antbird is one of a large family of approximately 200 species of mainly brown, shade-loving birds of the South and Central American forest understory. Seeing a name like "antbird" you would think their main source of sustenance must be obvious. You would be wrong. The family is named for following ants, not for eating ants.

Ants play a significant part in the ecology of tropical forests, not least the various species of army ants that rampage along the woodland floor eating every living thing in their path. These army ant columns are permanent fixtures; each individual army moves across an area during the day and forms temporary bivouacs at night. Due to the fact that there are always columns in the vicinity of any stretch of forest, some birds have made a career out of attending the ants, making their living by picking off the many insects that flee from the hordes. Panicked by the approach of the ants, these insects become very easy to catch.

Thus it is that, each day, the ocellated antbird wakes up and seeks out a column, especially one comprising the aggressive species *Eciton burchellii*. Checking bivouac sites and previously used paths, or listening for the calls of fellow ant followers, the bird will locate a column and stay with it throughout the day, fielding crickets, cockroaches and even such formidable prey as spiders and scorpions fleeing for their lives. The ocellated antbird rarely, if ever, hunts in any other way. With such bounty so readily available it doesn't need to.

Extreme

Families

Shortest courtship · Best speed dating · Most costly divorce · Sneakiest bigamist · Most relaxed attitude to breeding · Pushiest female · Hardest-working parents · Hardest-working male · Most frequent incest · Biggest chauvinist · Warmest nest · Coziest nest · Most eggs in a nest · Most eggs in a season · Largest extended family · Safest nest site · Most precarious nest site · Fairest chick provision · Cheekiest chick · Most expendable chicks · Most defensive nestlings · Largest day care · Bravest chick · Latest developer · Worst sibling rivalry · Keenest feminist · Best water carrier · Most suspicious mate · Strangest courtship · Canniest false alarm · Most notorious kidnappers · Most flagrant egg vandalism · Fiercest defense of nest · Dirtiest defense

There seems to be little time for idle decision making in the courtship of the great white pelican. Despite the fact that these birds have a potentially long breeding season to get things right, everything seems to proceed at an almost indecorous pace. It is no exaggeration to say that an individual can wake up in the morning completely fancy free and go to sleep that same night newly paired and with a nest site chosen and decorated. No other birds commit themselves so quickly.

In truth, pelican courtship is poorly understood. What is known is that the males first collect in groups and make loud mooing noises, intermittently pointing their bills skyward and using them in desultory jousts with other males. This attracts the attention of unpaired females, who then walk past the groups of males, causing the latter to follow behind as if joining a conga. After further calling and head-up posturing, the number of males following behind a given female gradually dwindles until finally just a single male is left in train. Incredibly this rather pallid act of persistence seems to earn the male his mate; the pairing process is evidently complete.

As if to proclaim its triumph, the pair proceeds to perform a display. The birds strut across the ground, holding their wings out for a few strides, while the male also waggles his pouch. After this things again move rapidly. Almost immediately both birds join part of the colony known as a laying group. Then they choose a nest site, their status as breeding birds earning them the right to keep this patch of ground for as long as they need it. If all this seems a trifle unromantic, the truth is that it doesn't last. The pair only persists for a single breeding attempt, and both birds acquire new partners a few months later.

Shortest courtship

NAME **great white pelican** *Pelecanus onocrotalus*
LOCATION Africa and Eurasia
BEHAVIOR engaging in a short but not very sweet courtship

These helmeted guineafowl might be finding time for a drink, but during their frenetic breeding season just about everything seems to be geared to speed and rush. As a result the male, in particular, may end the season exhausted, having lost as much as 11 percent of his body weight in just a few weeks.

It all starts when the nonbreeding flocks disintegrate as the sap rises and the males begin to get aggressive toward each other. Sometimes the tension explodes into a straight fight, but more often it involves what might be termed aggressive chasing. One guineafowl approaches another and ruffles the feathers on his back. This provokes the other male to start chasing him, as if the first bird had come up and mouthed an insult. The odd thing about the chase, however, is that the aggrieved male has no intention of catching his provocateur; he just keeps chasing on and on and on. The chasing is contagious, and, within a few moments, other males join in. Indeed, it is possible to have as many as eight birds running around in single file for minutes on end with no one bird ever "winning."

These chases are, meanwhile, carefully monitored by the females, presumably to assess the males on their sheer stamina. Before long, their interest sets in train another bout of frenetic activity, which could be likened to the curious human habit of speed dating. For several weeks males and females enter into a series of short-term relationships, one after the other, forming pairs as fluid and unpredictable as some couples in the human world. However, eventually more serious pair-bonds form and the guineafowl start to breed—that is, if the hard-working males have any energy left.

Best speed dating

NAME **helmeted guineafowl** *Numida meleagris*

LOCATION Africa

BEHAVIOR forming a succession of short-term relationships

Most costly divorce

NAME	**waved albatross** *Phoebastria irrorata*
LOCATION	Galapagos Islands
BEHAVIOR	forming long-term pair-bonds and paying a heavy price if they break

All long-term relationships have their ups and downs. It's true of people, and, to some extent, it is true of birds too. Recent research into albatross, for example, has shown that when things do go wrong in a relationship the consequences are dire, especially in regard to the reproductive potential of the partners.

Albatross are well known for their faithfulness. Divorce is almost unknown in the family; indeed, in one colony of wandering albatrosses (*Diomedea exulans*) it was recorded in a mere 0.3 percent of breeding pairs. Part of the reason for this could be explained by the fact that the unpaired birds are initially choosy. Young birds meeting for the first time won't rush into anything but will spend at least a season at the colony before attempting to breed the following year. They are very long-lived birds, so they don't want to be repenting at leisure, so to speak.

However, divorce does happen. It seems to be triggered by only one thing, and that is a series of failed attempts to breed. Such failures could suggest a lack of compatibility, or infertility, and in such situations a divorce may be the only way to make good the years before barrenness. Nevertheless it is a major step to take. Any bird divorcing cannot expect to breed the following year; that will be taken up by a year of careful courtship. And in some albatrosses a divorce can deprive an individual of 10–20 percent of its lifetime reproductive potential.

Sneakiest bigamist

NAME **European pied flycatcher**
Ficedula hypoleuca
LOCATION Europe, migrating to tropical Africa
BEHAVIOR two-timing

A male European pied flycatcher waits to deliver a caterpillar to its brood. At first sight it would appear to be similar to any number of other hard-working parent birds. This flycatcher, however, is harboring a secret. In spring the male arrives on the breeding grounds a week or two before the females and sets up a territory, defending it by singing. If all goes well he attracts a mate and breeding begins. The female lays a clutch of 4–7 eggs, and, as is typical among small birds, the female carries out incubation alone, which lasts 13–15 days. Once the eggs have hatched both parents share the hard work of feeding the nestlings.

Among a small but persistent minority of male pied flycatchers, however, this straightforward pattern conceals a less chivalrous mode of behavior. These birds take advantage of the fallow incubation period—when they have a couple of weeks off while the female is committed to the nest—to form another relationship with a different female. By the time the first female's brood begins to hatch the male has fathered a second clutch, which he now abandons completely, leaving the secondary female to bring up her clutch unaided.

It is not unusual for birds of many species to be bigamous, but what is very unusual about the pied flycatcher is the potential for deception. Among most species where it occurs polygamy is an open secret, in which two females share a single territory and each is surely aware of her rival. In the case of the pied flycatcher, however, the male holds two territories that can be some distance (up to 1¼ miles/2 km) apart in a different part of the wood. Thus it is possible that not only does the primary female not know about her mate's other partner, but the secondary female may never have known about the first.

These snow petrels might appear to be bickering, but at least they are attending their breeding site. That isn't always the case for a species that seems more accustomed than most to taking time off. The idea of taking a "sabbatical" from breeding might be unthinkable for most birds, but for seabirds it is less unusual. Most species of albatross, shearwater and petrel will sometimes do so, and some pairs will not even attend the colony if they themselves are not in suitable condition or if there is not enough food to make the effort worthwhile. Among snow petrels, whose feeding resources are linked to the amount of sea ice, a given individual will breed in only 52 percent of the seasons available to it during its lifetime.

This kind of schedule is a far cry from the frantic breeding seasons of small birds in temperate regions. Many birds, such as sparrows, tits and finches, have a life expectancy of just a couple of years, so they cram as much reproductive effort into their short lives as possible. By contrast, with no more than 7 percent adult mortality a year, a snow petrel can expect to live for at least 20 years.

Furthermore, snow petrels take a very long time to mature in the first place. Although some birds begin to attend colonies for learning purposes at the age of three, the majority of individuals wait until they are nine before they actually breed. It might seem wasteful—idle, even—but it's all a question of reproductive potential spread over a whole lifetime.

Most relaxed attitude to breeding

NAME **snow petrel** *Pagodroma nivea*
LOCATION Antarctic and subantarctic waters
BEHAVIOR taking a year out from breeding

Pushiest female

NAME **northern cardinal**
Cardinalis cardinalis
LOCATION North America
BEHAVIOR female sings instructions to male

Compared with the gaudy, brilliantly colored male—a glorious red all over—the female northern cardinal is drab. It does appear, however, that she wears the pants.

Among small songbirds it is highly unusual for a female to sing, but northern cardinals are an exception. The females sing less frequently than the males, but their songs are at least as rich, if not slightly richer, and they also learn them more quickly. If a male cardinal sings the female is perfectly capable of matching his song exactly.

Northern cardinal males sing a lot during the breeding season to keep their territorial boundaries intact. This carries on well into the nestling stage, when the male is also spending time looking for food for the young. It is about this time that, on frequent occasions, the female seems to "interrupt" him.

Careful studies have shown that there can be a subtle duet between male and female, which appears to have the function of instructing the male when to come to the well-hidden nest with food. Normally the female simply lets the male sing away, but on occasion she will add a few phrases of her own. If these phrases match those of the male nothing much happens; but if she sings a non-matching phrase the male is highly likely to come down with food. It appears that the non-matching phrase is the male's cue to come and provide. When the female calls, the male comes running.

A blue tit delivers a caterpillar to one of its youngsters in the nest. This is a snapshot of a single event in a very long day for this particular parent; a record of one of more than 500 visits made between dawn and dusk. The other parent will do the same, which probably makes blue tits the hardest-working parents of any bird species in the world. During the most intense period of feeding, when the chicks are about nine days old, the adults deliver, on average, 97 caterpillars to each youngster per day. A pair in England was even recorded as managing 1,083 feeds on one busy day in May.

Why should the blue tit condemn itself to such hard labor? There are three main reasons. Firstly the days of late spring in the Northern Hemisphere are very long, allowing plenty of time to notch up such an impressive tally of feeds. Secondly blue tits time their breeding to coincide with a highly seasonal glut of food and thus lay more eggs than might be expected in a single brood, rather than a moderate number stretched over several broods. And the third reason is the nature of the food.

You might expect a parent on a feeding expedition to catch more than a single item of food, but the blue tit carries only one item at a time back to the nest. The reason is that caterpillars can bite and injure nestling blue tits, so the parent breaks the jaws of its prey and removes any noxious hairs or fluids. The preparation needed allows the bird to work on only one caterpillar at a time in a remarkable demonstration of parental devotion.

Hardest-working parents

NAME **blue tit** *Cyanistes caeruleus*
LOCATION Europe and western Asia
BEHAVIOR making 1,000 visits a day to the nest

Hardest-working male

NAME **northern harrier** *Circus cyaneus*
LOCATION widespread throughout the Northern Hemisphere
BEHAVIOR keeping multiple partners and young

A male northern harrier flies along holding his latest prize. It is the breeding season, and this bird is about to deliver the prey to his mate waiting on the nest. Nesting harriers, along with many birds of prey, practice a very strict code of conduct in the summer, in which male birds are the sole providers for an incubating female and, later, for the chicks in the nest.

This can be a huge burden. The male will begin provisioning the female intermittently before the eggs are laid, which may involve a few feeds a day. Once incubation is underway, however, his labors begin in earnest. The incubation period lasts for about a month, during which the female may not leave the nest at all. Things get worse when the eggs hatch. The average clutch is 4–6 eggs, and, in the early days, the male has to provide for all the chicks, bringing in a minimum of about six large kills a day. It is not until a week later that the female at last contributes to the hunting effort, but even then she regularly checks the nest and will cover the young in inclement weather, making her efforts somewhat feeble compared to those of the male.

If you think this is hard work, consider the fact that some males are polygamous, with three or four nests on the go at once. In theory a male could be providing for itself, four females and 24 young. In practice, however, this does not happen. Nests experience losses, and males have favorites, so while some nests are well provided for others will receive few provisions.

Most frequent incest

NAME **purple swamphen** *Porphyrio porphyrio*
LOCATION New Zealand
BEHAVIOR frequent interfamily copulation

Relationships can be adversely affected when you live in suffocatingly close quarters. This seems to be the conclusion from a study concerning an isolated population of the purple swamphen, a bird found through much of the Old World and Australasia. Over most of its wide range the purple swamphen forms monogamous pairs and leads a conventional family life, but in New Zealand, where the lack of habitat restricts the space available to the birds, things are very different. With nowhere to go the pukekos, as they are known locally, are forced into groups that are either stable and based around families or unstable, violent and unproductive.

The situation among the family groups is especially intriguing. Groups consist of one or two females, up to seven males and an assortment of nonbreeding helpers, made up of the progeny from previous years. Within the arrangement, it seems, just about every adult mates with everybody else—and that really does mean everybody. Homosexual interactions, which are rare in birds generally, are regular. Even more frequent is incest. With just about every bird in the group related to each other it is probably unavoidable. At least, however, every member of the group, including the nonbreeders, contributes to the care of young in the communal nest.

In the nonfamily groups cooperation is replaced by competition. Males guard their mates assiduously, trying but often failing to keep rivals away. The act of copulation is continually interrupted and successful breeding is often a rarity. Not an ideal arrangement either way.

At first sight this male knobbed hornbill looks as though he is using his huge bill to micromanage a morsel of food for his own consumption. Take a closer look below his leaning figure, however, and you will see a hole in the tree. Inside, the bird's mate is sitting on her eggs, waiting to be fed.

But this is no conventional food delivery. The female is "inside" in more senses than one: she is effectively imprisoned. A couple of weeks earlier she will have entered a suitable hornbill-sized cavity and made herself at home. Then, little by little, her entrance hole will have been plugged by excreta and mud, until her only contact with the outside world is via a small vertical slit through which she can defecate and peer. She is unable to fend for herself and now depends solely on handouts from the male.

Superficially this looks like a case of abuse and the worst kind of male chauvinism. But in fact the female, together with her clutch, is shut in for safety reasons until the eggs have hatched and the chicks are ready to fledge. Hornbills are large birds, and any suitable treetop hole is likely to be vulnerable to attacks from snakes, monkeys, martens and other tree-borne predators. Narrowing the entrance to a slit keeps most of these intruders out.

So not only is the female complicit in her incarceration, she actually plugs the hole herself, using her bill as a trowel. Furthermore, the imprisoning arrangement is tough on the male, who faces weeks of hard work bringing food to the family. In truth, this chauvinist is, in fact, a perfect gentleman.

Biggest chauvinist

NAME **knobbed hornbill** *Aceros cassidix*
LOCATION Sulawesi, Indonesia
BEHAVIOR "imprisoning" the female inside the nest

Warmest nest

NAME **common eider** *Somateria mollissima*
LOCATION circumpolar northern coasts
BEHAVIOR building a nest with exceptional insulation

The eggs seen here are among the best looked after of any in the avian world. There are two reasons for this. First of all they are incubated assiduously by a female with great staying power. The eider duck sits for 25–28 days and rarely leaves her nest, making minimal efforts to feed herself. If enemies approach she will abandon the eggs only when the danger is upon her and all has been lost. Otherwise she remains passive, allowing her camouflage to keep the clutch safe from predators.

Secondly these eggs are bedded in the finest down in the world. To say that the clutch lies in luxury might be overstating the reality of life on these chilly northern coasts, but the down that the eider plucks from her own breast to insulate her nest is regarded by humans as a natural product of immense value. At current prices a complete quilt filled with eider down fetches about $11,000 (£6,000). Iceland still exports 5,500 pounds (2,500 kg) of down a year, farmed from eiders that breed on coastal properties where they are carefully protected from predators.

The down is the best insulating material known, superior to the down of other wildfowl or birds in general and to all synthetic materials. Apparently the clusters of eider down have unusually dense cores, where the inner parts of the feathers intertwine and trap air. Down is measured in "fill power" — the volume, in cubic inches, filled by an ounce of down. In eider down this can reach as high as 700 cubic inches. A single ounce (28 g) contains 2 million interlocking filaments, making it the softest, lightest and warmest of all.

Coziest nest

NAME **American goldfinch** *Carduelis tristis*
LOCATION North America
BEHAVIOR building a snug, watertight nest

A female American goldfinch brings in food for her young. In contrast to many small birds, this species doesn't bring in juicy caterpillars, worms or other small or soft-bodied invertebrates; instead it provides seeds regurgitated in a kind of paste. Very few songbirds feed their young on such "difficult" food.

There are several other unusual features in the domestic life of the American goldfinch. For a temperate songbird, for example, its breeding season starts exceptionally late. It is mid-July before most goldfinches start to raise young. This may be for several reasons. American goldfinches molt in the spring, which might make the multitasking load of changing feathers, nest building and laying eggs too much for the adults to bear. The birds may also be awaiting the midsummer blooming of thistles, their favorite food. At any rate, many goldfinches lay eggs as late as August, with the young fledging in October.

Another remarkable feature is the nest itself. Built mainly by the female, although the male may bring in material, it is a snug cup constructed from various plant fibers woven together exceptionally tightly with spiderwebs and plant down (especially thistledown). This nest is placed in the fork of a tree, 33 feet (10 m) up, and is so perfectly constructed that, later on when the young have left, it can hold water as reliably as any human-made receptacle.

Most eggs in a nest

NAME **ostrich** *Struthio camelus*
LOCATION sub-Saharan Africa
BEHAVIOR producing a large and variable clutch of eggs

The nest being tended by this ostrich in South Africa contains a healthy tally of eggs, but it's a long way off the record for the species. Some nests hold 20–25 eggs, and the world record is a mind-boggling 78. That's a lot of young ostriches! But not all of these will be the product of a single bird. Most nests host the eggs of at least 2–5 females, with a record 18 individuals contributing to a single clutch. You could, therefore, call these communal nests, but ostrich politics are not as egalitarian as that term implies. Not all the eggs are equally valued.

Each ostrich nest is tended by the territory-holding male and one female, who is known as the "major hen." She acquires this status by being the first female to lay an egg at the nest site. Over a period of two weeks or so the major hen lays about eight eggs in the same nest; meanwhile other females are invited to lay eggs too. These secondary females, known as "minor hens," mate promiscuously with other local males and contribute nothing to incubation anywhere, despite laying eggs in several nests.

This leads to the question: why do the male and major hen tolerate the presence of extra eggs in their nest when they get no assistance and have no genetic stake in the eggs? The answer appears to be that they use the extra eggs as a buffer against predators. If an enemy strikes the sheer numbers of eggs will reduce the chances of the major hen's eggs being destroyed. Furthermore, the major hen usually recognizes her own eggs and makes sure that they are at the center of the clutch, thus ensuring they will be incubated. The eggs from the minor hens, meanwhile, are arranged in a circle around the nest. Many will not be incubated, but some will—ensuring that there is some value in the arrangement for the minor hens as well.

Most eggs in a season

NAME **brown-headed cowbird**
Molothrus ater
LOCATION North America
BEHAVIOR laying more eggs than any other wild bird

The domestic scene in this photograph appears idyllic enough, but all is not as it seems. If you look closely you will see that one of the chicks has a yellow gape, while the rest have a pink gape. That's because the yellow-gaped chick is not the offspring of the attending wood thrush (*Hylocichla mustelina*), but is actually an interloper. During the laying of the clutch, a brown-headed cowbird stole in and laid its own egg, destroying one of its host's in the process. Now, unwittingly, the wood thrush is raising a cowbird along with her own chicks.

The very same scene is replicated in untold numbers of nests every breeding season in North America. The brown-headed cowbird is a generalized brood parasite, which means that it will lay its eggs in the nests of many different species, not just wood thrushes. In fact, in all it has been known to parasitize more than 220 different species and, of these, 144 have successfully raised chicks contracted out to them. Being a larger species than the cowbird, this wood thrush is in a way fortunate. If a cowbird lays in the nest of a species smaller than herself her young will aggressively outcompete the host youngsters, causing them all to starve to death.

Another extraordinary feature of the brown-headed cowbird's bizarre life is the fecundity of the female. With so many possible hosts to attack, she just carries on laying eggs for as long as she can. In the wild it is not uncommon for the female to lay 40 eggs in a season, and in captivity a figure of 77 has been recorded. (The common cuckoo, *Cuculus canorus*, by comparison, lays only 20.) The downside is that it causes a great deal of destruction and reduced production among the cowbird's legion of host species.

Largest extended family

NAME **noisy miner** *Manorina melanocephala*
LOCATION eastern Australia
BEHAVIOR up to 22 birds feeding the young in one nest

Of all the world's young birds, few seem to have more attention lavished upon them than the nestlings of one of eastern Australia's most common and overbearing birds, the noisy miner. The youngsters may be fed up to 50 times an hour, one of the highest rates known among birds.

Furthermore, this is very much a collective effort. An extraordinary total of 22 adults has been recorded visiting a single brood during the nestling stage to bring edible gifts, with a range of 6–21 being recorded on other occasions. All this for a clutch of only 2–4 eggs on average. These young are treated like royalty.

The extraordinary attentiveness of the adult members of this extended "family" is partly due to the fact that the noisy miner is a colonial, cooperatively breeding species. When the female is nest building, which takes about a week, she continually advertises what she is doing with a special display so that all the males in the area know what is going on and when the young might hatch from their eggs. Why they should all help her, though, is unclear. Although the birds behave promiscuously, most clutches are sired by just one father.

Noisy miners have an astonishingly complicated social structure. Males live within their specific "activity spaces," often overlapping with those of other males. Birds with overlapping spaces form long-lasting associations known as "coteries," usually comprising 10–25 birds.

Coteries make up extended "colonies," consisting of hundreds of birds, and sometimes the coteries divide into short-lived gangs known as "coalitions." It is generally the males in coteries that, together, help at a single nest—which might partially explain why so many birds are compelled to engage in this useful, though curious, type of feeding activity.

Safest nest site

NAME **great dusky swift** *Cypseloides senex*
LOCATION central South America
BEHAVIOR nesting behind a waterfall

Just below the rim of the thundering Iguassu Falls, on the border between Argentina and Brazil, four great dusky swifts appear to be toying with their lives. Wheeling inches from the roaring water, the slightest misjudgment of their flight path could cause them to strike the fast-moving surface and be washed away to certain death.

But it's an everyday risk the swifts take because, to them, the benefits of living near the waterfall greatly outweigh the dangers. Indeed, paradoxically, it is sanctuary they seek in this extreme environment. By placing their nests beside the falls, or even behind them, the swifts can keep their eggs or young uniquely safe. No predator in its right mind is likely to climb down the perilously moist rocks, bathed in permanent spray, to reach the narrow ledges upon which the swifts have built their nests. And even if a predator wanted to do so, the barrier of falling water would certainly prove impenetrable.

There are a few disadvantages, of course, to living in such an environment. The saturated conditions mean that the breeding adults have to cope with spray on their plumage when incubating or brooding. The young, on the other hand, have a generous coating of down to keep them warm. Aside from the inherent danger, the highly aerial swifts have all that they need. Luxuriant plant growth on the nearby rocks provides nesting material, and the swifts have a short commute to nearby forest to catch insects for their food.

Most precarious nest site

NAME **white tern** *Gygis alba*
LOCATION tropical islands
BEHAVIOR laying its egg on a narrow branch

On first seeing this picture you are probably wondering what that egg is doing there. There's no nest in sight—just an egg balancing on a branch with an ethereal-looking white seabird staring soothingly at it. Something is surely not right.

But it is. Despite the incongruity of it all, this egg is lying where it was laid, put there intentionally. What you are looking at is the nest site of the white tern, perhaps the most precarious nest site of any bird. The slightest of bumps would dislodge this egg irretrievably, causing it to tumble down to the sand below where it would probably crack. Yet the tern risks this balancing act for safety's sake, to protect its egg from ground predators.

Remarkably this site is relatively low for a white tern. It is common for the eggs to be in the canopy of trees, sometimes a dizzying 66 feet (20 m) above ground, although they are usually at about half that height. The branch seen here is, admittedly, on the narrow side. Most eggs are balanced on typically horizontal limbs that measure at least 4–5 inches (10–13 cm) wide, but only 3 inches (8 cm) has been recorded.

Terns don't leave everything to chance. They normally scratch out bark with their feet to give the egg more balance, and they will also lay the egg in a natural depression. Furthermore, the incubating adults are careful to approach the egg from behind during a changeover and to leave the site by falling backward. Even so, accidents and strong winds lead to high wastage.

The adults incubate the egg for 28–32 days and then the youngster hatches. Fitted with strong claws, a chick can hold onto a branch rather better than an egg can, but the 60 days that pass before it fledges must still be a nervous time for all.

Fairest chick provision

NAME **common kingfisher** *Alcedo atthis*
LOCATION much of Eurasia, Indonesia and North Africa
BEHAVIOR giving each youngster a fair chance to be fed

A common kingfisher tends its young. The kingfisher burrow might not look like an unusual scene but what goes on here is truly exceptional. Most parent birds don't instinctively look out for the welfare of every one of their chicks. The worst offenders only serve food preferentially to their firstborn, often allowing other chicks to starve, while at the majority of nests a delivery is something of a free-for-all, and the most vigorously begging nestlings receive first pick each time. A nest full of chicks is a competitive, ruthless environment.

Except, that is, among kingfishers. In the nesting chamber a rare type of fairness prevails. For their first two weeks nestling kingfishers wait to receive food in what is effectively a line. They lie down in a circle, back to back, and just one of the youngsters orients itself toward the light coming in from the nest-hole entrance. When an adult arrives with a fish only this one chick begs, and it alone is fed. The rest remain passive. Once a chick is fed it moves out of the way, and the next youngster takes its place.

This remarkable arrangement, known as the carousel system, ensures that each youngster is regularly fed. Furthermore, if a nestling tries to cheat the rest of the brood administers group justice, pecking at it and on occasion throwing it to the back of the chamber.

Cheekiest chick

NAME **herring gull** *Larus argentatus*
LOCATION Eurasia
BEHAVIOR abandonment of useless parents

In the birding world, tales abound about the tribulations of various hapless youngsters. They are variously abandoned, killed or starved by parents, murdered by their nest fellows, munched by predators or drowned. It is, therefore, a joy to recount the case of a chick prepared to take its own path in life, regardless of the perils besetting it on every side.

In the picture opposite, a young herring gull has just entered the world. It is a chick of the so-called "semi-precocial" type, covered with down but with its eyes open and with legs that enable it to stand and walk. Although motile it does not feed itself and does not tend to wander away from the borders of its parents' territory. That would be suicidal in a gull colony. Next door's adults won't take kindly to strange chicks running into their territories, and sometimes they can respond with savage violence.

Sooner or later a parent will arrive with a helping of food, regurgitating it in front of the chick. This is the youngster's lifeline, and it is upon such handouts that this cute, fluffy ball of helplessness depends for its survival.

But is this chick really as helpless as it looks? Not always, it seems. Herring gulls are famous for their chutzpah, and it starts early. Some chicks, believe it or not, monitor the handouts from their parents, even from the start, and come to the conclusion that their providers are not doing a good enough job. They make an instinctive decision that, if they stick with their lot, they are unlikely to survive for long—so they summarily dismiss their parents.

Such a decision involves taking an extraordinary risk. Yes, they have to slip next door and hope that either they will not be noticed or that they will somehow be accepted into the brood. Doubtless some fail and are evicted or killed, but for others the strategy clearly works. And if it does, it could be their passport to 30 years of productive life and, with luck, more effective parenthood.

Most expendable chicks

NAME **hooded grebe** *Podiceps gallardoi*
LOCATION Argentina and Chile
BEHAVIOR leaving its inconvenient chicks to die

An inhabitant of lakes on high-altitude plateaus in southern South America, the hooded grebe was only discovered in 1974. In many ways it seems incredible that it has survived at all, so slow and fitful is its reproduction rate.

The problem for the hooded grebe is that there is a conflict of interest between its nesting and feeding requirements. It only builds its nest on large, floating mats of a plant called water milfoil, or *Myriophyllum*, which grows out on the middle of the lake at a safe distance from the shore. But, being diving birds, the grebes find it extremely difficult to forage among the *Myriophyllum* because the plant is so dense. Hunting is therefore confined to a comparatively narrow stretch of open water between the waterweed and the shore.

To make matters worse, these lakes do not support fish, which are nutritious, but only small crustaceans, such as amphipods, and a few large snails. In order to get enough sustenance the adult grebes have to work unusually hard, making long dives up to 33 feet (10 m) deep throughout the daylight hours. In such circumstances the hooded grebes only raise one chick at a time, both parents directing their full efforts toward their single offspring. They actually lay two eggs, and both may hatch, but the younger of the two is merely an insurance policy against the nonviability of the first and is left to starve.

Furthermore, in the prebreeding season, before the eggs hatch, the adult birds sometimes exhaust the supply of large snails. In such cases they sometimes just abandon their surviving chick and move to a different lake, putting their own lives ahead of their unfortunate offspring.

Most defensive nestlings

NAME **hoopoe** *Upupa epops*
LOCATION Eurasia
BEHAVIOR nestlings using various strategies to keep predators away

A parent hoopoe arrives with a meal for its nestlings, approaching her nest with unusual confidence. Many an adult bird on feeding duty comes upon a scene of disaster, discovering that in its absence its young have been taken by a predator. But far from being helpless "lambs to the slaughter," young hoopoes are unusually resilient in the face of danger. Indeed, they can pose as much trouble to intruders as Macaulay Culkin's feisty character in the movie *Home Alone*.

The first defense used by the nestlings is an olfactory one, which is useful, since the majority of potential enemies are mammals with sensitive noses. Only four days after hatching the young develop a gland on their rump that produces a foul-smelling substance. This stench, which is also made by the adult female, may pervade the nest site. Upon reaching the noses of approaching mammals it will act as a deterrent, keeping the would-be predators away.

In addition to this, however, once they are six days old the young hoopoes acquire another defensive measure. This comprises a physical as well as an olfactory assault. If a carnivore is foolish enough to reach the nest the youngsters are able to shoot a jet of urine straight at it, which is just as disgusting as the smelly substance that pervades the nest. Amazingly, a given youngster can hit a predator from 24 inches (60 cm) away. If the intruder is at the entrance to the nest the chances are that a direct hit will result.

If that isn't enough the youngsters can try intimidation. Just in case the source of danger hasn't already got the message, the chicks are able to hiss like snakes. And if all else fails they have sharp bills, which they jab toward their enemy with all the gusto required. Not surprisingly, the fledgling survival rate in hoopoe nests is unusually high. Ask any bloody-nosed, urine-coated, foul-smelling predator running away from what it believes to be a snake.

Largest day care

NAME **greater flamingo** *Phoenicopterus ruber*
LOCATION Africa and Eurasia
BEHAVIOR forming enormous groups of chicks

An army of greater flamingo chicks occupies the shallows of a lake in Spain, the young's plumage a gray counterpoint to the adults' famous pink. Among the most sociable of all birds, flamingo young gather together in masses, just as they will do when they are adults. Greater flamingo "day cares" may comprise a few thousand birds; in the lesser flamingo (*Phoeniconaias minor*) of East Africa, however, 300,000 young may bunch together to form the largest concentrations of any young birds anywhere.

Flamingos lay only one egg and, when it hatches on the peculiar mud cup that acts as the nest, the chick grows quickly. Within 12 days (or as little as five if disturbed) it leaves its birthplace, already able to walk and swim, and joins a throng of peers. The youngsters do this for protection's sake, the sheer numbers reducing their odds of being singled out by a predator, and the masses offering some physical protection. At first they are guarded by a few adults, probably birds that have lost their own eggs, but within a few days there is only a single guard for every several hundred chicks.

In the day care, the young hang around until, at various points in the day or night, their respective parents come to feed them. Not surprisingly it can be difficult to find your progeny in a group of 300,000, but the flamingos seem to manage. The visiting adults probably know roughly where the chick is among the crowd, and the loud, individually recognizable begging call of the chick, learned in the intimacy of the nest, helps the parents locate their offspring.

Bravest chick

NAME **common murre**
Uria aalge
LOCATION circumpolar northern waters
BEHAVIOR jumping off a cliff

The young murres in this picture are a couple of weeks old, each living a comfortable life on a cliff top with its parents bringing it about 1 ounce (28 g) of food every day. It has no siblings—no common murre ever does—so there are no contests over food. All these two have to do is to put on weight and hang around. For now they haven't got much to worry about. In a week's time, however, each faces a huge ordeal and the key to a new world. The young murres must make the transition from cliff top to ocean—by jumping!

At first sight these two quarter-grown chicks seem hopelessly unprepared for leaping from a cliff that could be nearly 985 feet (300 m) high. After all, they cannot fly and their wings have not yet grown beyond the primary coverts. So when they jump all the youngsters can do is flutter their stubby winglets, exerting minimal control and adding only a slight brake to their fall. But they are also extremely light and fluffy and can glide a bit. Although some chicks undoubtedly come to grief, their chances are nonetheless good.

Naturally, though, there is an anxious air to a murre colony on the nights the youngsters jump, the adults calling loudly and pleadingly to their progeny as they line up to go from the best ledges. Most jumpers depart at dusk, perhaps because their many enemies, mainly gulls and jaegers, might be slightly less abundant and active at this time. Their descent is quick, but some predators still manage to intercept them, either in midair or on the surface of the sea.

When a youngster has reached the water it quickly joins its father, using individually recognizable call notes. Once united, the two birds swim hurriedly out to sea where the parent bird feeds his offspring for several weeks in the relative safety of the ocean.

Latest developer

NAME **wandering albatross** *Diomedea exulans*
LOCATION southern oceans
BEHAVIOR not beginning to breed until over 10 years old

A lot of water will have passed under this young albatross's wings before it emulates its attending parent and joins the ranks of the breeding. Of all the world's birds, albatross take the longest to attain the age and experience needed to set about the costly—in terms of energy—and potentially risky business of breeding.

According to extensive studies, male wandering albatrosses start to breed, on average, when they are 10.7 years old, as opposed 10.4 years for the females. Some individuals wait until they are 13. If the latter is the case, current estimates suggest that such a bird will have flown 1.5 million miles (2.4 million km) before staring its reproductive life.

Why should albatross begin to breed so late? There are several reasons, one of which is life expectancy. When a bird can live to 30 years or more and has to replicate itself only once it can afford to start late. Secondly it takes many years for an albatross to become proficient enough at finding food for it to take on the extra task of providing for a chick. Thirdly albatross take a long time to establish their lifelong pair-bond. And fourthly various social dynamics, such as the availability of a mate or nest site, might prevent a bird from starting early.

Physiologically, wandering albatross could start earlier than they usually do. Males at five years have the same levels of testosterone in their blood as do breeding adults, as well as adequately developed testes. The females, on the other hand, have smaller ovarian follicles up to the age of eight compared with those of breeding adults, and their hormonal balance suggests that they are physiologically incapable of breeding until that age. Thus, the actual age of first breeding is slightly later than it could be in reality.

Incidentally, the record for the oldest debutant breeding albatross is actually held by a relative, the light-mantled sooty albatross (*Phoebetria palpebrata*). One individual was recorded as having begun at no less than 16 years old.

Worst sibling rivalry

NAME	**lesser spotted eagle**
	Aquila pomarina
LOCATION	Eurasia
BEHAVIOR	the killing of one chick by another

The bird in this picture has something in common with just about every other adult lesser spotted eagle, besides being of the same species. Overwhelmingly it will have been the firstborn in its nest. Younger siblings have a survival rate so low as to be negligible, and their fate is not a pleasant one.

Along with most large eagles, lesser spotted eagles usually lay two eggs. The first egg tends to be laid three days ahead of the second and, because it is incubated as soon as it appears, it hatches three days ahead as well. Thus, right from the start, the firstborn is larger and heavier than its sibling.

Lesser spotted eagles probably lay two eggs as insurance against the first not hatching. In doing so, however, they are unleashing a terrible and uneven conflict between the two siblings. Both chicks are exceedingly aggressive toward each other, but the firstborn, with its greater age and weight, always has the upper hand. Its attacks, which involve plucking the down of its sibling, pecking at its head or shaking it, may not in themselves cause life-threatening injuries, but they do have the effect of intimidating the younger bird into submission. The latter retreats to the corner of the nest and dies, either from starvation or exposure or due to falling out of the nest.

So certain is the demise of the second-born chick that this phenomenon is known as the Cain and Abel conflict. It results in enormous losses every year; if, as has been estimated, there are 50,000 breeding pairs of the lesser spotted eagle in the world, then about 40,000 second-born chicks may die annually.

Keenest feminist

NAME **western gull** *Larus occidentalis*
LOCATION western seaboard of North America
BEHAVIOR rearing chicks without a father

A parent western gull feeds her chick by regurgitation in a scene repeated across thousands of gull colonies all over the world. It is a delightful snapshot of the intimate workings of gull society.

There may be nothing unconventional about this chick's family, but some western gull chicks are known to grow up in a very unconventional setup indeed—one that is virtually unknown among birds other than a few gulls. These chicks grow up as the offspring of "married" couples that comprise two females. The females function exactly as a pair, with each partner contributing to nest building, one or both females laying eggs, and both tending the chicks until they are fledged. To all appearances it is just another nest.

These pairings can produce young because each partner goes to a nearby male when she requires sperm for her eggs. The female-female pairs do not form out of homosexual orientation, however, but usually arise because of a lack of males in the population. Neighboring males are invariably content to copulate and donate their sperm, because this will contribute to their own personal productivity, but their commitments and efforts lie elsewhere.

The arrangement is far from ideal, however. Measurement of nesting success has shown that, on the whole, the rate of hatching equates to only half that achieved by conventional pairs, an anomaly caused by a higher incidence of egg infertility. But being imperfect does not mean that the practice is so peripheral as to be exceptional. In one well-studied colony, 10 percent of all pairings were between two females. In fact, it is actually a practical adaptation in colonies where the male-female ratio is skewed, and females may revert to heterosexual pairings in future years.

A male Burchell's sandgrouse visits a waterhole in Botswana. It is a daily ritual for these birds because their diet of seeds compels them to drink regularly; each dawn sees them taking commuting flights to a reliable water source. This bird appears to be taking things a step further by having a bath at the same time.

It's a very special type of bath, though, almost unique to the sandgrouse family. What this male is doing is getting his belly feathers so completely soaked that, in a few moments, he can fly back to the nest with his feathers loaded with water. His chicks are thirsty, and, once the male arrives, they will gather under his belly to have a drink of what remains on his plumage "sponge."

In order to carry out this water delivery the male sandgrouse has specially adapted feathers on his belly. The microscopic projections on the feathers' branches—the barbules—are usually coiled and lie flush against the vane (the flat plane of the feather). When soaked, however, the barbules uncoil and project at right angles to the vane, forming a bed of hairs that absorbs water by capillary action. The result is an absorption rate superior to a sponge; a male sandgrouse is estimated to be able to carry up to 1⅓ fluid ounces (40 ml) of water. Some, however, is lost to evaporation during the flight home.

Best water carrier

NAME **Burchell's sandgrouse** *Pterocles burchelli*
LOCATION southern Africa
BEHAVIOR carrying water soaked in its belly feathers

Most suspicious mate

NAME **great tit** *Parus major*
LOCATION Eurasia
BEHAVIOR escorting the female all day long

Two great tits have a skirmish at a bathing site. It is spring, and the males are bundles of nerves. Tension is boiling over.

It is the season when every male great tit must be unusually vigilant. From the period of about two weeks before laying eggs to the time, a month later, when the clutch is finally complete, a female great tit is highly liable to stray and compromise her mate's treasured paternity. Although great tits are monogamous, females regularly seek extra-pair copulation to add variety to the genetic material in the nest. It is up to their mates to stop them.

The lengths males go to to guard their mates are remarkable. Their primary weapon is song; every day throughout this period, at dawn chorus time, the male chooses to perch right outside the nest holes and sing. Once awake and active, the females come outside and, with the male ready and waiting, the sexes copulate.

During the day the male takes the same stifling approach. He keeps the female company at all times, fighting off unwelcome suitors (as seen here) and ensuring that nothing happens without his knowledge. It is during the evening, perhaps, that his paternity is most at risk because most eggs are laid at dawn. A slack moment would court disaster. So the male redoubles his efforts until, at last, the female goes to roost in the nest, still escorted by her mate.

Then, just to make sure, the male stays outside the nest, a chaperone by the door. He might wait 15 minutes before, confident of the female's sleepy state, he himself can at last retire to roost.

Strangest courtship

NAME **western bowerbird**
Chlamydera guttata
LOCATION western and central Australia
BEHAVIOR doing more than most to impress the females

Pity this poor western bowerbird—or indeed any male bowerbird. Few, if any, species in the world have to work harder to win the affections of the opposite sex. In most bird species a bit of singing and display goes a long way, but with bowerbirds that's only half the story. Bowerbirds do indeed sing and display, but they also go to astonishing lengths to make a special construction to impress the females. This construction—known as a bower and built purely for display—involves the male in hour upon hour of painstaking work followed by virtually continuous maintenance all year round. The quality of workmanship is one of the crucial factors in a female's choice of potential father to her chicks.

The bower of the western bowerbird is known as an avenue bower. It consists of two parallel stick walls, each about 14 inches (36 cm) long and 9 inches (23 cm) high and standing 6 inches (16 cm) apart. At each end the male clears away debris on the ground, replacing it with a large number of decorative objects, in this instance some green fruit; on some bowers there may be as many as 200 of these items. The designer also adds other objects, which may include bleached bones, snail shells, pebbles and sometimes human artifacts, such as glass, metal and even spent shotgun cartridges.

As if that isn't enough, the bower needs a little interior decoration as well. For this reason the bowerbird selects a juicy fruit and daubs the pulp on the inside walls, as if adding a coat of paint. But females are notoriously hard to please, and, even after years of work, some males fail to father any chicks.

Canniest false alarm

NAME **barn swallow** *Hirundo rustica*
LOCATION throughout much of the world
BEHAVIOR sounding a false alarm

Barn swallows, especially those that live in colonies, are notoriously prone to erring from the path of conjugal fidelity. Both males and females, although secured in a formalized relationship with a member of the opposite sex, nonetheless frequently find time to consort with others and thus exchange genetic material.

To a female an extra-pair copulation may be an advantage, since it offers sperm that could be of superior quality to that of her social mate. To a male, however, the idea that his mate would stray outside the pair-bond is an anathema—it would compromise his paternity. Thus male swallows regularly undertake "mate guarding." It is well named: the male follows his partner everywhere while she is fertile, ensuring that she does not have the opportunity to consort with a stranger.

Mate guarding, however, isn't easy. Sometimes the male loses sight of his mate, dramatically increasing the potential threat to his paternity. So what does he do? He gives an alarm call, warning the colony of attack from a predator. Everybody interrupts what they are doing and flies up into the air, bunching together, calling.

Except that there's no attack. It's a deliberate false alarm. In the tightly packed melee, however, it is much easier for the male to find his mate. A little deception goes a long way—especially when it keeps the female faithful.

Most notorious kidnappers

NAME **white-winged chough** *Corcorax melanorhamphos*
LOCATION eastern Australia
BEHAVIOR kidnapping the fledglings of neighbors

One of Australia's strangest birds, the white-winged chough is an inhabitant of the dry country of the east. It lives in small groups of about seven birds, which forage by walking in lines in a sinister fashion, like policemen conducting a precise search.

Each group is a close-knit, permanent association, consisting of a dominant male, several mature females and a few immature birds, the latter offspring from previous seasons. In the breeding season the group defends a communal territory, and tasks, such as constructing an unusual mud-bowl nest, brooding the chicks and bringing the nestlings food, are shared by everybody. It seems that bringing up youngsters is difficult for white-winged choughs, so much so that it requires at least four adults to complete the job effectively. Not surprisingly, recruitment into the group is painfully slow.

If the group is small, however, white-winged choughs have a way of circumventing the problem of too few helpers. They pick a fight with a neighboring group and kidnap a fledging from them. During the frequent melees between clans adults have been seen displaying at a fledgling, enticing it to swap flocks.

The kidnapped fledglings don't always remain with their "captors," but often they do, and for the long term. As a result the genetic variability within a group is enhanced, and the birds have that vital extra help on hand next time they attempt to breed.

When thinking about the kinds of birds that might destroy eggs in a nest, the chances are that smaller species, such as the cactus wren, may never occur to you. Cuckoos, gulls and birds of prey, maybe—they look mean and have a reputation to match. But a cute little wren?

Well, yes. Wrens are, believe it or not, notorious for their habit of breaking the eggs of other birds—not only of rivals among their own kind, but also the clutches of unrelated species. The cactus wren, for example, regularly sneaks into the nests of a small desert bird known as the verdin (*Auriparus flaviceps*); so much so that the nesting success of the latter can be seriously impaired on a local level. Once at a nest the wren pricks all the eggs in a frightening act of intentional vandalism.

The cactus wren is by no means the only offender. In North America the northern house wren (*Troglodytes aedon*) has been known to attack the nests of neighboring chickadees and sparrows, while the marsh wren (*Cistothorus palustris*) has a virulently fractious relationship with the red-winged blackbird (*Agelaius phoeniceus*), a much larger species than itself.

But why do they do it? In general the eggs are not eaten, so this is unlikely to be a significant reason. In terms of cavity-nesting birds, such as the house wren, it could be to ensure a long-term supply of nest sites. The cactus wren probably does it to wipe out competition for resources from neighboring birds. But the truth is, this widespread habit among wrens, but not other small birds, is still not fully explained.

Most flagrant egg vandalism

NAME **cactus wren** *Campylorhynchus brunneicapillus*
LOCATION deserts of North America
BEHAVIOR destroying the eggs of other birds

Fiercest defense of nest

NAME **snowy owl** *Bubo scandiacus*
LOCATION tundra of the Northern Hemisphere
BEHAVIOR attacking human intruders

Owing to its broad head, cuddly looks, clean-sheet white plumage and friendly reputation peddled by movies and books, the snowy owl is a popular bird. But if you come across one on the tundra of North America or Eurasia, don't be deceived by its docile expression. This owl is big and, on occasion, exceedingly aggressive. It won't kill you, but it can still seriously injure you.

The snowy is one of a fearsome triumvirate of northern owls that really should be avoided during the breeding season; the others are the great gray owl (*Strix nebulosa*) and the Ural owl (*Strix uralensis*), the latter being found only in Eurasia. These owls can be utterly fearless. They regularly attack dangerous predators, such as Arctic foxes and northern goshawks (*Accipiter gentilis*), and the great gray has been known to try to defend its nest site from a bear.

The owls are far from gratuitous in their attacks on people; they are only aggressive when they have young in the nest, and some pairs are less excitable than others. But that doesn't mean they are not effective. Their talons are, of course, sharp enough to cut your scalp or face and potentially blind you.

Furthermore, the birds almost always attempt to attack from behind and, with their silent wings, can take you completely by surprise. A snowy owl might attempt to brush you with its wings or back, but the other two species concentrate on your head and shoulders. And owl attacks can be persistent. One bird watcher noted that dive-bombing may last at least five minutes from first encounter and may go on for 20 or more.

Because these owls may not breed at all in years when rodent populations are low, meaning that nesting sites are at a premium, it is understandable why they should attack a dangerous intruder with such vigor. But that does not make them any less intimidating—or potentially dangerous.

Dirtiest defense

NAME **fieldfare** *Turdus pilaris*
LOCATION northern Eurasia
BEHAVIOR dive-bombing nest predators, with a difference

In the winter there is little to distinguish the fieldfare in behavior from any other member of the thrush family. It feeds on fruit and invertebrates and lives in loosely organized flocks. On the breeding grounds, however, it engages in a remarkable routine that is so far unrecorded in any other bird.

The fieldfare's sociability extends into the breeding season, the birds usually breeding in large aggregations that are a little too widely spaced to be called colonies. However, when the nests are threatened by a predator, there is a strong neighborhood solidarity that resolves into an impressive defense for the sake of collective security. This only happens at the crux of the breeding season, when young are in the nest.

If a crow enters the fieldfares' settlement, with the likelihood of causing trouble, the reaction is immediate. The adult fieldfares fly up into the air, and, one by one, they give out a loud screech, dive down toward the predator and deliver a bomb of feces upon it. If enough individuals take part the unwanted parcels will rain down upon the enemy and begin to stick to its plumage. It must now retreat for its life. If too much feces clogs the feathers the plumage may be damaged, and, in extreme cases, the bird will be unable to fly and will eventually die.

It is not surprising, then, that many other species of bird place their nests in the vicinity of fieldfare settlements in preference to elsewhere. With that kind of protection available it makes an unusually safe place to raise a family.

Index

Page numbers in *italic* refer to illustrations